Copernicus Books
Sparking Curiosity and Explaining the World

Drawing inspiration from their Renaissance namesake, Copernicus books revolve around scientific curiosity and discovery. Authored by experts from around the world, our books strive to break down barriers and make scientific knowledge more accessible to the public, tackling modern concepts and technologies in a nontechnical and engaging way. Copernicus books are always written with the lay reader in mind, offering introductory forays into different fields to show how the world of science is transforming our daily lives. From astronomy to medicine, business to biology, you will find herein an enriching collection of literature that answers your questions and inspires you to ask even more.

Dirk Huylebrouck

Dark and Bright Mathematics

Hidden Harmony in Art, History and Culture

Birkhäuser

Dirk Huylebrouck
University of Leuven
Brussels, Ghent, Belgium

ISSN 2731-8982   ISSN 2731-8990 (electronic)
Copernicus Books
ISBN 978-3-031-36254-5   ISBN 978-3-031-36255-2 (eBook)
https://doi.org/10.1007/978-3-031-36255-2

Translation from the Dutch language editions: "Lugubere 'wiskunde'" and "Hemel en aarde in de 'wiskunde'" by Dirk Huylebrouck, © Maklu-Garant 2021 and 2023. All Rights Reserved.

© The Editor(s) (if applicable) and The Author(s), under exclusive license to Springer Nature Switzerland AG 2023

This work is subject to copyright. All rights are solely and exclusively licensed by the Publisher, whether the whole or part of the material is concerned, specifically the rights of reprinting, reuse of illustrations, recitation, broadcasting, reproduction on microfilms or in any other physical way, and transmission or information storage and retrieval, electronic adaptation, computer software, or by similar or dissimilar methodology now known or hereafter developed.
The use of general descriptive names, registered names, trademarks, service marks, etc. in this publication does not imply, even in the absence of a specific statement, that such names are exempt from the relevant protective laws and regulations and therefore free for general use.
The publisher, the authors, and the editors are safe to assume that the advice and information in this book are believed to be true and accurate at the date of publication. Neither the publisher nor the authors or the editors give a warranty, expressed or implied, with respect to the material contained herein or for any errors or omissions that may have been made. The publisher remains neutral with regard to jurisdictional claims in published maps and institutional affiliations.

This book is published under the imprint Birkhäuser, www.birkhauser-science.com by the registered company Springer Nature Switzerland AG
The registered company address is: Gewerbestrasse 11, 6330 Cham, Switzerland

Are there more things between heaven and earth, as Hamlet surmised, than can be grasped by man? In this book the double answer is yes, there are more, but no, they are understandable, as mathematics keeps them together.

Em. Prof. Dr. Jean Paul Van Bendegem, Centre for Logic and Philosophy of Science, VUB, Belgium.

Foreword

I spent much of the first half of my life as a visual artist and craftsperson. I created 3D sculptures from a variety of materials, especially blown glass. I always had an affinity for recreational mathematics and magic, learning a lot from articles in Scientific American from the great Martin Gardner. But these were always just side interests for me, for fun, and I did not see a connection to my primary interest of art.

My life changed when I became the primary caregiver for my son, Erik. Suddenly, my main concern was to learn about his interests. I started homeschooling him from grade 2 and loved the challenge of figuring out how to adapt material to what would draw his attention.

When Erik was seven years old, he showed an interest in computer programming, motivated by a curiosity of how video games get made. I learned a few commands in the BASIC programming language: PRINT, GOTO, and STOP. Using just these commands on a neighbor's PC, we made our first video game, a simple text adventure game. Erik was hooked and learned programming far faster than I could keep up, but I kept him fed with books to learn from. (This was before the Internet.) A few years later, he had consumed what books could teach him, so I suggested he sit in some university classes. He loved them, and soon (at age 12) he was officially enrolled and taking classes as fast as possible. I sat in with him on the classes so I could still be able to communicate with him about the taught subjects.

When Erik became a Ph.D. student (at age 15), he fell in love with computational geometry in two and three dimensions. This intrigued me because I could see a connection to my background in visual art. My self-taught 3D visualization skills seemed useful, so we started working together on research. Now, the roles were reversed, and Erik taught me about the underlying mathematics, while I could imagine and invent directions to explore. The first problem we worked was in mathematical magic—folding paper and making one straight cut to produce a complex shape—which I had read about in Martin Gardner's column years before.

Flash forward 11 years: in 2007, Erik and I are an art/math/science team at MIT, making research and art together, and we are on our first sabbatical in Brussels, Belgium. It was a great adventure, with wonderful people, architecture, and food.

One of the wonderful people we met was Dirk Huylebrouck. I recall during an early meeting explaining to Dirk about how stimulating Brussels was. During one of our first meetings, I mentioned one of our unusual art ideas. Dirk has the gift of getting people to open up and talk about who they really are. I mentioned we were fascinated by the cobblestone streets in Brussels and were considering using chalk to create a pattern on the streets that could only be seen from the air. I remember his smile from the conversation and was blown away by Dirk's ability to listen to people, not be judgmental of crazy ideas, and in fact become intrigued and engaged in other's ideas.

This book represents a lifetime of listening and understanding people and their ideas. Dirk not only relates to their mathematical understanding and achievements, but he also describes a why, and who they are as intriguing and interesting people.

While this is already an impressive feat for people he meets (such as Erik and me), he has even achieved this with historic mathematicians. How does he do it? I conjecture that he has the ability to time travel and visits these people during their lifetimes. It sounds implausible, but I think he is able to close his eyes and imagine who they really are.

As you read the book, I had like you to consider a puzzle. This tome is a collection of independent articles published in EOS. You can read the chapters in any order. The puzzle is this: What was the original order the articles were written? And what is the statement Dirk is creating by the specific order they appear in the book?

All in all, this book is an incredible achievement that spans many years. It will inspire you to understand not just the ideas behind some incredible mathematics, but how people get intrigued, understand, and solve problems.

Martin Demaine
Angelika and Barton Weller
Artist-in-Residence
Massachusetts Institute
of Technology
Cambridge, MA, USA

Introduction

This is a collection of some of the author's contributions written for the science magazine EOS, a Dutch equivalent of Scientific American, distributed in Belgium and The Netherlands. This monthly journal collaborates with Scientific American, translating some of its papers, so the comparison makes sense. Having been published previously in a popular science magazine such as EOS, these chapters are proven to be accessible for a broad audience.

The book is designed in such a way that every chapter can be read independently, which means that a subject will be repeated rather than referred to. For instance, the golden section, a frequently recurring topic in general mathematics, appears several times in this book. The references are limited, as popular science magazines believe they obscure the page image and include them only if there is a special need. In addition, today it is easy for the reader to look up information using the keywords mentioned in the text.

It is hoped that the reader will appreciate that all the chapters that follow have an original point of view and are not just adapted copies of what anyone can read on the Internet. Sometimes, the content is a little provocative, but hopefully that is an invitation for further reading. Likewise, the subdivision of this book is perhaps surprising. Sure, to many, mathematics seems to come from hell, but this part in fact refers to lugubrious stories about math as it relates to skulls, murders, or World War II. The more down-to-earth part is about how math might illuminate maps, money, Facebook, folding paper, shapes in ice or some down-to-earth as yet unsolved math problem. 'Heavenly math' alludes to Vedic, Islam and New Age math, to a meta-divine section, and concludes with an interview with a top mathematician who also wrote about the existence of God.

Admittedly, the title of the book is but a common denominator to group together different stories from different countries and different periods in time, while the math in this book may be no 'real' math for die-hard mathematicians, because it is sometimes misapplied in the examples that are discussed. The geographical, cultural, and historical diversity herein will perhaps inspire those who want to illustrate mathematics with surprising examples from the real world.

Special thanks go to Ch. Bartlett, Em. Prof. at the Towson University Art Department, for proofreading a draft version of this book.

Contents

Part I Math from Hell

1 Hell, Earth and Heaven in One Painting 3
  Holbein's Deformed Skull 4
  Myth in the Museum 5
  Through a Glass 7
  Hungarian Escher 9

2 Hitler's Math 13
  Counting with the Nazis 14
  National Socialist Math Problems 14
  Propaganda Issues 16
  New Order Math Problems 16
  War Mathematics 17
  Current Issues 17
  Pamphlets with Issues 18
  Post Scriptum 20

3 Guernica 21
  Picasso's 1937 Masterpiece 22
  Picasso on the Warpath 23
  Geometry and Guernica 24
  Endless Tinkering 25
  To This Day 27

Contents

4	**Architect-Alchemist**	29
	Racist Proportions	30
	Modulor	32
	'Mathematics' by Le Corbusier	34
	Deliberately Wrong	36
	Golden Ratio Buildings	38
5	**War Hero, Math Genius, Martyr**	41
	Ultimate Recognition for Gay Icon and Top Scientist Alan Turing	42
	Life and Work	42
	The Turing Test	45
	Mathematical Nuance	46
	A Tragic Death	47
	Repentance Comes After Sin	48
6	**Murder and Higher Math**	51
	Prisoners Find Their Passion in Numbers and Formulas	52
	Food, Math, Sleep	53
	Havens' Continued Fractions	54
	'Held Back' from Congresses	55
7	**Murdering Emperors**	59
	Mathematical Intrigues	60
	Survival Statistics	61
	Power Curves	63
	Meaningfulness	65
	Applicable Today?	66
8	**When the Dead Talk in Code**	69
	After the Zodiac Killer and Langrenus, an Altar Code?	70
	Cryptography or Abbreviations?	71
	Echo of the Merovingians	74
	Magritte's Surrealism	74
	Contemporary Opinion	77

Part II Down-to-Earth Math

9	**Columbus's Reference Line on Earth**	81
	Georg von Peuerbach	82
	Oradea Today	83
	Each His Own Meridian	85

10	**Mathematical Stock Advice**	89
	Financial Astrology	90
	Holy Geometry	90
	Stock Markets	91
	Stock Geometry	92
	Fake Predictions	95
11	**The Fall of the Lottery**	99
	Coincidence or not?	100
	There's Something Going on	102
	Misled Intuition	104
	A Statistical Case?	106
12	**The Mozarts of Mathematics**	109
	From Greyhound to MIT	110
	Magic and Origami	112
	Playful Mathematics	114
	Mathematical Improvisation	115
	Light-Hearted	116
	Update	118
13	**Cold and Austere Beauty in Harbin, China**	123
	An Icy Collaboration	124
	Hyperboloids and Cones	125
	Ice Construction Research	127
14	**Your Friends Are More Popular**	131
	The Math of Facebook and Twitter (X)	132
	Scale-Free Networks	134
	Power of Algorithms	136
15	**The Most Down-To-Earth Problem**	139
	Mathematics with a Warning	140
	Vain Attempts	141
	Tao Tried His Luck	143
	Second Opinion	144
16	**Meccano Math**	147
	Nobel Meccano	148
	Meccano Polygons	149
	Hands-On Mathematics	151

Part III Math from Heaven

17 Mathematical Meditation — 157
A Vedic Math Hype — 158
Magical Mathematics — 160
More Sutras — 161
Modern Fantasies — 164

18 Did Newton's Apple Fall First in India? — 167
Did Jesuits Import Calculus from India? — 168
Mathematics from Kerala — 170
What Went Wrong? — 172

19 Allah's Nonagons — 175
Mathematical Embellishments — 176
Surprise: Nonagons — 176
Mystery — 181

20 Is Mathematics Halal? — 183
Mathematics is Evil — 184
Bad Grades — 184
A Great Past — 186
Mathematics Becomes Meaningless… — 188
…And Meaningful Again — 189

21 The Church's Perspective — 191
Perspective and Parallel Projection — 192
A Mixed Form: The Fishbone Representation — 195
Reverse Perspective — 198
Mathematical Saints — 201

22 The Flower of Life — 205
New Age Math — 206
Tetrahedra — 209
Spatial Stacking — 212

23 A Meta-Divine Nautilus — 215
A Beautiful Shell — 216
Shell Statistics — 216
A Meta-Divine Ratio — 218
Similarities with the Divine Number — 220
A New Myth? — 221

24	**Happiness in Unprovability**	225
	An Interview with Harvey Friedman	226
	About His Work	227
	A 'Simple' Unprovable Mathematical Statement	228
	Miscellaneous Questions	229
References		231

Part I

Math from Hell

1

Hell, Earth and Heaven in One Painting

The deformed skull in a painting by Hans Holbein becomes more visible when you look through a glass. His anamorphosis still inspires us, even today.

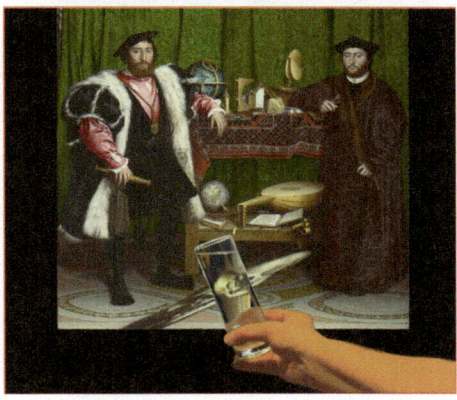

Fig. 1.1 Hans Holbein painted a deformed skull on his work *The Ambassadors* (1533), which becomes more visible through a glass

© The Author(s), under exclusive license to Springer Nature Switzerland AG 2023
D. Huylebrouck, *Dark and Bright Mathematics*, Copernicus Books,
https://doi.org/10.1007/978-3-031-36255-2_1

Holbein's Deformed Skull

In 1533, Hans Holbein the Younger, a German Renaissance painter, made the oil painting *The Ambassadors* (Fig. 1.1). The two-by-two-meter work presents a double portrait of Jean de Dinteville, an ambassador of the French king, and Georges de Selve, a French Bishop. The painting is on display in London's National Gallery.

What is special about this work is not so much the depicted persons, but the objects and their symbolism. The deformed skull at the bottom draws the attention of the spectator. It has been reproduced in numerous publications. For example, the painting adorns the cover of the book *Disharmony of the Spheres* in which Jennifer Nelson (Penn State University Press, 2021) describes the Europe of Holbein's Ambassadors. Holbein's skilfulness illustrates the use of perspective, which was not only a leap forward for art—the Flemish Primitives got their name from their ignorance about vanishing points and horizons—but also opened a new field in mathematics. It flourishes to this day, under the name of 'projective geometry'.

The visual mystery of *The Ambassadors* can be properly solved in only one way. One must look at the painting from the centre of the right side to the bottom left corner. Because of the perspective effect, the left side is reduced (because it is far away) and so one suddenly sees a skull in the correct proportions (Figs. 1.2 and 1.3).

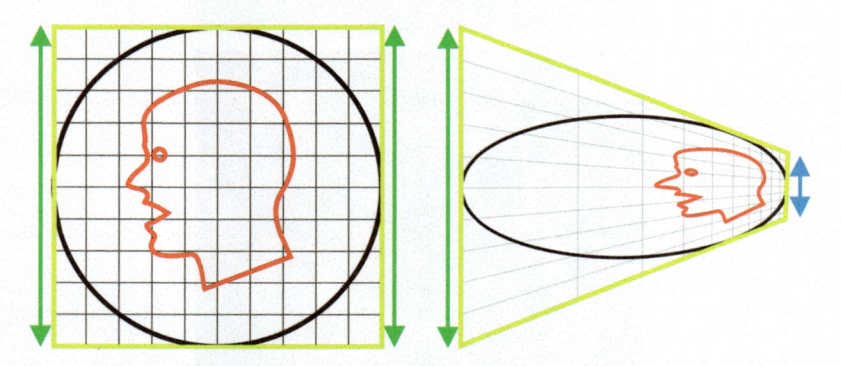

Fig. 1.2 The original drawing of a skull inscribed by a circle (left) and the same drawing after a perspective transformation as in Holbein's painting (right)

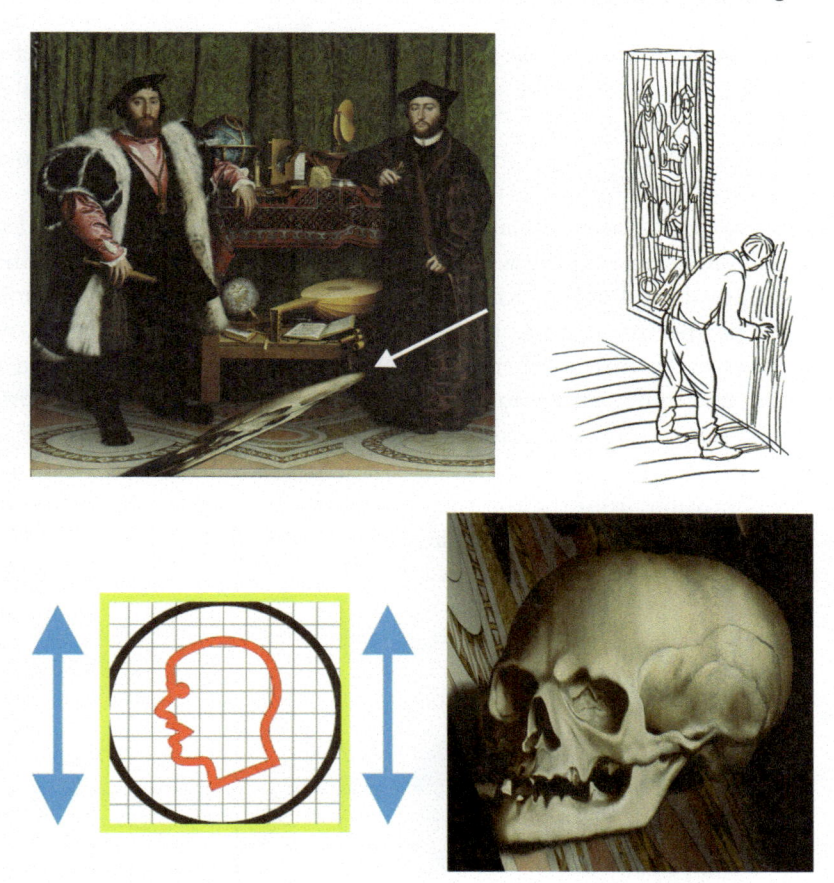

Fig. 1.3 The correct viewing direction recreating the skull in its original proportions

Myth in the Museum

A persistent myth has it that the skull can be viewed equally well from both the right and the left. Yet, if one looks from the bottom left to the top right, one will see a very distorted skull. The myth may be explained by the fact that a circle turns into an ellipse after a perspective transformation. The image so distorted can turn again into a circle when viewed from the left and the right. However, a skull in such a circle will be distorted even more if it is viewed twice from the left.

This was food for a second myth: the painting should be viewed from the bottom of a staircase. London's National Gallery even made a computer animation reconstructing the image of the skull from a left and from a right viewpoint. It showed how the painting should be placed: on an intermediate landing of two flights of stairs, so that it could be viewed from above (which gives the correct image of the skull) and from below (which gives a very distorted image). The museum sold the program with the bug to the Microsoft Art Gallery (Fig. 1.4).

When London's National Gallery had Holbein's painting restored and was thinking about relocating the work to a staircase, mathematician John Sharp (1945–2020) of the London Knowledge Lab revealed the misinterpretations about the perspective distortion (Fig. 1.5). He wrote about it in the official pages of the British mathematics website and published a paper in the *Art Review* magazine. These publications got the attention of *The Guardian* newspaper, but the museum remained convinced that one could see the painting correctly from both the bottom left and the top right, when in truth this is only possible from the top right. Officially, it did not want to respond to the criticism, but it did admit that it was wrong: the painting was finally installed away from all staircases, just above the ground in room C, no longer viewable from the bottom left.

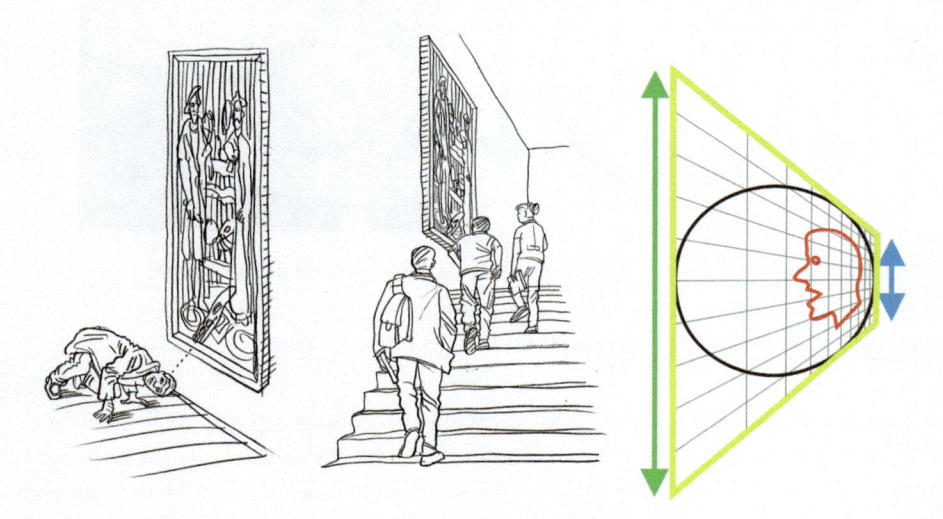

Fig. 1.4 A perspective transformation of the skull viewed from the left, can change the ellipse again in a perfect circle, as would be the case if the painting hung at the top of a staircase, but the skull would remain deformed

Fig. 1.5 John Sharp in front of the London National Gallery

Through a Glass

But why did Holbein want to depict a deformed skull, and moreover, so prominently on the painting? One theory is that in this way the two ambassadors are placed between three 'levels': the heavens, represented by the astrolabe and other astronomical objects on the top shelf; the terrestrial world depicted by the globe, the books, and the musical instruments on the bottom shelf; and the underworld, represented by the deformed skull under the feet of the ambassadors. When one looks at the painting through a glass, the shape of the skull becomes visible, and the three layers are placed one below the other.

The explanation seems correct, but a mathematical problem arises. If, for the sake of simplicity, a cylindrical glass is used, then looking through that glass will never turn an ellipse into a nice circle, but into an angular oval, because of the effects in the cylindrical lens. This is illustrated, for example, in an anamorphosis on the painting *Edward VI* (1546) by William Scrots. Depending on the distance between the painting, the lens and the observer, the image is also left–right mirrored (Fig. 1.6).

Fig. 1.6 Looking at the painting through a glass puts the heavens, the earth and underworld together (left), but a glass also distorts the image, as in a painting by W. Scrots (right)

Even without the additional lens effects, one quickly sees that proportions contract differently when using a perspective transformation instead of simply squeezing a circle into an ellipse. In the first case, distances shorten rapidly as they approach the horizon, while in the second case they are all reduced in the same way. In the reverse method, rectangles are all stretched in the same way, without taking into account the perspective distortion. This is clearly visible on Holbein's skull: the eye cavities become different in size and the jawbone becomes too long. This could not have been the intention of a precision painter like Holbein, who represented all other objects almost photographically (Fig. 1.7).

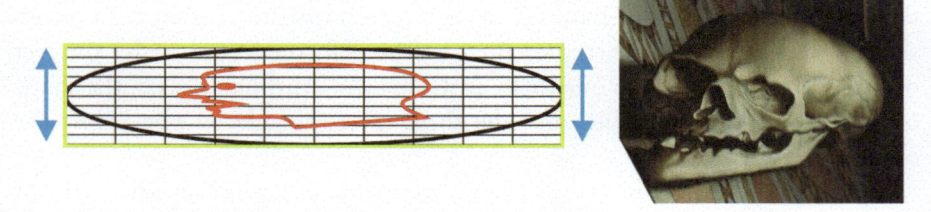

Fig. 1.7 Squeezing a circle into a narrow ellipse distorts squares differently than through the perspective method (left) and this has consequences for the reconstruction of Holbein's skull (right)

Hungarian Escher

Holbein's anamorphosis inspires us, even today. The Hungarian graphic artist István Orosz, for example, made dozens of drawings with 'hidden skulls' and is a specialist in anamorphoses. Sometimes referred to as the 'Hungarian Escher', the artist has a long-standing reputation in mathematical art circles, through lectures at conferences, exhibitions, and animated films with graphical representations of mathematical paradoxes. At one of these conferences on mathematics and art, he also discussed Holbein's painting, about which he wrote a book (*A Követ És A Fáraó*, The Ambassadors and the Pharaoh, Typotex Publishing, 2011) (Figs. 1.8 and 1.9).

After a visit to Belgium, Orosz also tried out a beer glass of the brand 'Duvel', inspired by the distortions in Holbein's painting and those obtained by looking through a glass. A Duvel beer glass is almost spherical at the bottom, and so the image is inverted horizontally as well as vertically (Fig. 1.10).

Fig. 1.8 István Orosz at his desk at the University of Sopron (Hungary)

Fig. 1.9 Hidden skulls by István Orosz

Fig. 1.10 A drawing by István Orosz and its distorted image through a Duvel beer glass

The image through the almost spherical Duvel beer glass is reminiscent of the glass sphere in the painting *Salvator Mundi*, which was attributed to Leonardo da Vinci in 2011. In this painting, the Christ figure holds a glass sphere in his left hand, but the image seen through that sphere is not inverted and appears more like the image that would be seen through a flat circle of glass. This error could prove that the painting is not by Leonardo da Vinci, as some consider him an infallible genius. Conversely, in view of the many errors da Vinci made in other paintings, designs and machines, this may be a good indication that the work is indeed by his hand (Fig. 1.11).

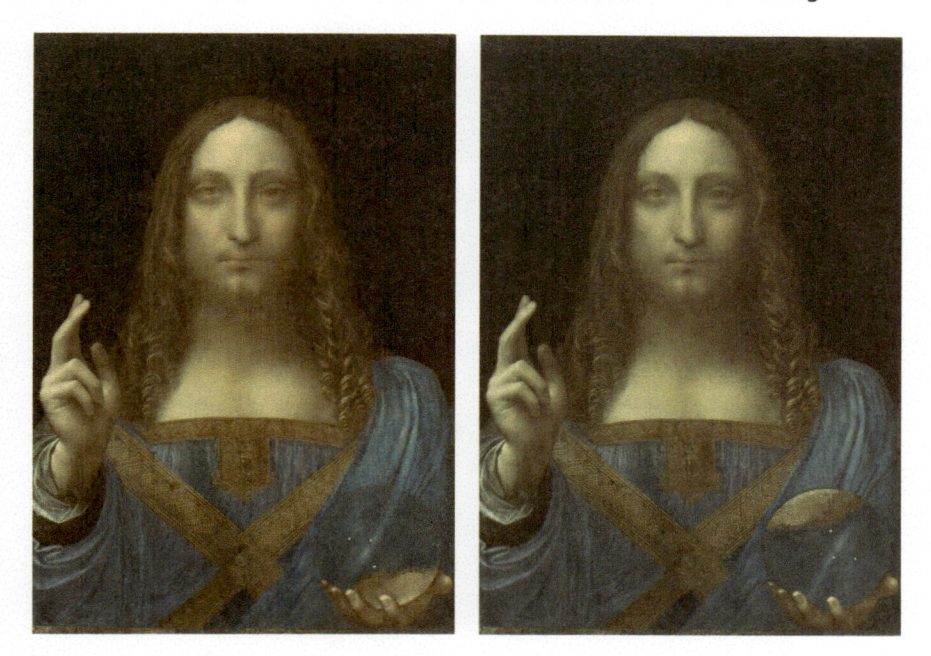

Fig. 1.11 The painting *Salvator Mundi*, attributed to Leonardo da Vinci (left), and the same work with the corrected image through the glass sphere (right)

2

Hitler's Math

A teacher using a swastika in geometry class is guaranteed to cause trouble. In Nazi Germany, ideologues infiltrated mathematics lessons.

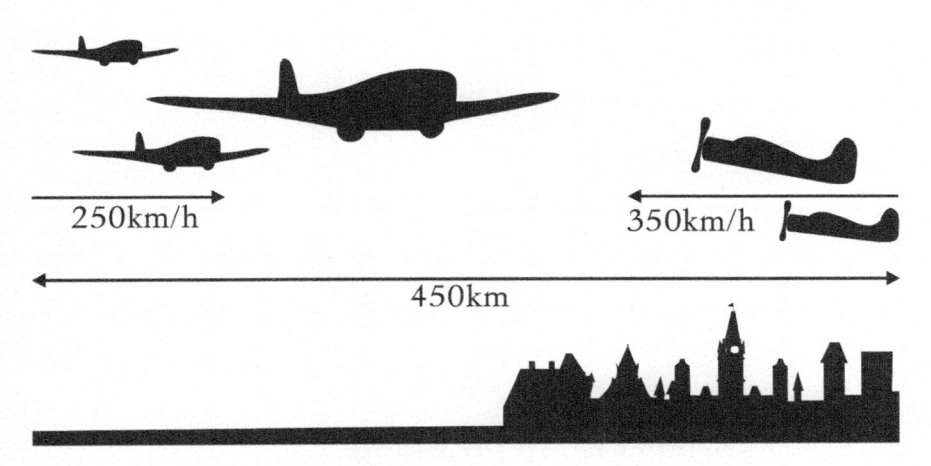

Fig. 2.1 A typical Nazi textbook question. An enemy bomb squadron is approaching a city at 250 km/h. While it is still 450 km from the city, a squadron of high-speed fighter planes takes off at 350 km/h to defend the city. At what distance from the city does the battle begin? And when do the squadrons clash?

Counting with the Nazis

Writing about Hitler is challenging. In Belgium, an episode of a famous cooking show was not broadcast because it was about Hitler's favourite trout dish. Yet, a radio program about Hitler's preferred angora rabbits was permitted. Hitler bred them during the war.

Talking about Hitler in a math class is risky too. A French teacher from Saint-Clément de Rivière near Montpellier was reprimanded for using a swastika in geometry class. She wanted to make a connection with a talk about World War II a student had given just before her class. The incident even made it to the national television channel TF1.

How the Nazi ideology infiltrated German education in the times before and during WWII can be easily imagined for most courses. In language lessons, German was, of course, presented as the most beautiful language, while the so-called Aryan heritage was taught in history lessons, and the territorial demands of the Nazis was on the geography curriculum. The incident with the French teacher mentioned above was a misunderstanding, but her idea was not that strange: during the Nazi period, German pupils were given exercises in geometry based on Nazi symbols.

A certain Adolf Dorner managed to complete an entire booklet entitled *Mathematik im Dienste der nationalpolitischen Erziehung* (Mathematics in the service of national political education). It appeared in 1936. Later, Gerhard Kölling and Eugen Löffler added three more volumes 'for further study', the *Mathematisches Unterrichtswerk für höhere Lehranstalten* (Mathematical educational work for higher education). Below you can read some of the mathematical problems the Hitler Youth had to study, divided into categories. They did not sprout from some wild neo-Nazi fantasy, but literally come from Dorner's book (Fig. 2.1). The last problem though is from another book, *Hitler Youth, 1922–1945, An Illustrated History*, by Jean-Denis G.G. Lepage.

National Socialist Math Problems

1. Draw a swastika in the square badge of the Hitler Youth so that the width of the black stripe is related to the width of the white stripe as s: w. What are the most common values for s and w? (Fig. 2.2)

    In a circle with a diameter of 8 cm, draw a swastika, of which the black bars are the same size as the white spaces and are 1 cm wide (Fig. 2.3).

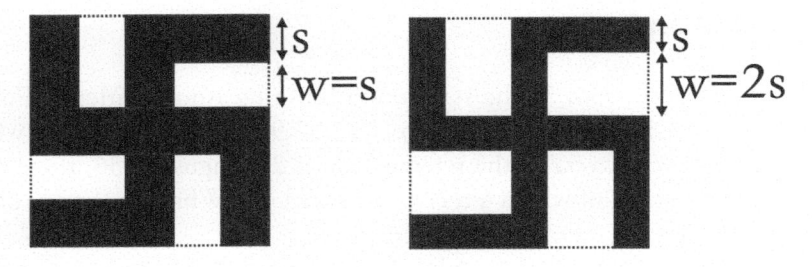

Fig. 2.2 Two values for *s* and *w*

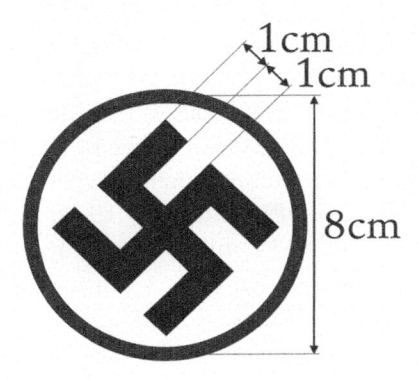

Fig. 2.3 The requested drawing

2. a. A Jungzug (a battalion of the Hitler Youth) has 71 Pimpfen (the name for recruits). In the first two groups there are 15 young people, in the third 14 and in the fourth 13 Pimpfen. How many are there in the fifth group?
   b. Four Fähnlein (a medieval unit for the size of a battalion, also used by the Hitler Youth) of a group in the Hitler Youth number 145, 148, 140 and 142 boys. How strong is the group?
   c. An Untergau (a subgroup) of Jungmädel (young girls) has 2780 members, one shire numbering 590 girls, the second 572, the third 498, while the last two are equally strong. How strong are they?
3. How much space is required for a Kameradschaft with 15 boys, a Schor formed by three Kameradschaften and a Gefolgschaft (three Scar), if they march in a column in rows of three? The distance from the centre of one body to the centre of another body from right to left should be 70 cm, and between the front and back 105 cm.

Propaganda Issues

1. On 19 August 1934, in the referendum electing Adolf Hitler as Führer and Reich Chancellor, there were 42,695,908 valid votes, of which 38,395,479 voted yes. In the referendum of 29 August 1936, there were 44,954,937 votes, of which 44,411,911 voted yes. What percentage of the vote was this?
2. In the years 1938 and 1939, Adolf Hitler united 97.55 million Germans of blood and race: the 67.06 million of the Altreich (the Old Empire); the 6.20 million of the Alpine and Danube shires (Austria); the 3.50 million from Sudetenland (the Czech border area); the 0.20 million from the Protectorate of Bohemia and Moravia (the rest of the Czech Republic); and the 0.12 million from Memelland, a border region with Lithuania. How many inhabitants did the Great-German Empire have in 1939?
3. In 1931 there were 4.355 million unemployed in the German Empire, 5.103 million in 1932, 3.849 million in 1933, 2.282 million in 1934, 1.714 million in 1935, 1.035 million in 1936, and 0.469 million in 1937 (each time on 30 September). Compute the annual decline in unemployment between 1932 and 1937.

New Order Math Problems

1. According to calculations, a psychiatric patient costs the state around 1500 Reichsmark (RM) every year, a Hilfsschüler, a pupil from a pedagogical school, 300 RM, a Volksschüler or pupil from a primary school 100 RM, while a student from middle or higher education costs about 250 RM. Present the amounts in a bar chart.
2. According to conservative estimates, there are 300,000 psychiatric patients, people with epilepsy, and so on, in German healthcare facilities. How much do they cost per year at a daily rate of 4 RM? How many wedding loans of 1000 RM can the government issue each year with that money?
3. The population (in millions) of the three main groups in Europe changed between 1900 and 1930: Germanic peoples 124–149; Roman peoples 103–121; Slavic peoples 166–226. Calculate the growth factors and the growth figures for ten years for the three groups under the assumption of uniform growth. What would the proportions of the three population groups be like in 1960, based on the same growth? For the three dates, calculate the proportions in hundredths of the three ethnic groups

in relation to the total population of Europe. What serious danger do you perceive for the future of the Germanic peoples if fundamental change does not occur? Fortunately, there is a legitimate hope of a reversal in population growth in Germany.

War Mathematics

1. An enemy bomb squadron is approaching a city at 250 km/h. While it is still 450 km from the city, a squadron of high-speed fighter planes takes off at 350 km/h to defend the city. At what distance from the city does the battle begin? And when do the squadrons clash? (Fig. 2.1)
2. A 1-ton chemical bomb is 70% chemicals. How many bombs of this type are needed for the contamination of Berlin, an area of 2 by 2, if an area of 1 by 1 requires 20,000 kg of phosgene? How many aircraft should you use if each aircraft can carry three such bombs?
3. An airplane flies at a speed of 240 km per hour to its target 210 kms away to drop bombs. When can you expect it back if the bombing takes 7.5 min?

Current Issues

Another question from Adolf Dorner's list concerned the 'angular profile' of a skull. This angle is formed on the one hand by the 'German horizontal', the plane through the lower edge of the eye socket and the highest point of the external auditory canal, and on the other hand the nose, from the bridge of the nose to the height of the upper jaw edge. As a geometry exercise, students were asked to determine the profile angles of skulls from photographs. Note that there were special protractors so that one could determine whether someone had a Jewish or Aryan nose.

Artist Ruth Mateus-Berr, professor at the University of Applied Arts in Vienna, is one of the rare Austrians who does not shy away from the dark past of her country. With her 2009 work *Vermessen*, she referred to social Darwinism used in Nazi Germany as a pseudoscientific justification for racism. Today, it is a characteristic of extreme right theories. One of her works is a photograph that proposes to use a colour chart in which colours are converted into numbers, to decide from what value on someone can be called black or white (Fig. 2.4).

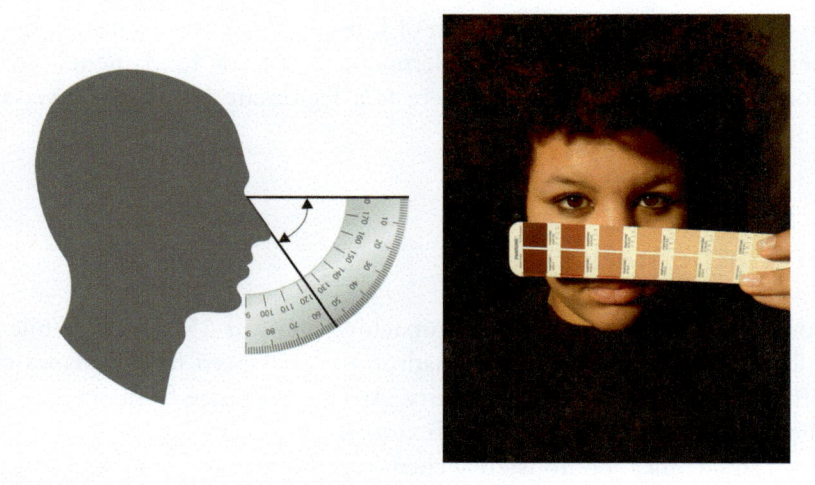

Fig. 2.4 Nazis determined whether someone was Jewish or not using the angle of the nose (left); artist Ruth Mateus-Berr proposed a colour chart to find out how brown someone is (2010, right)

Take former US President Barack Obama, who is 50% black and 50% white. Yet everyone called him a 'black president'. If his wife Michele had been white, were their children white or black? The question therefore arises: from which brown values is someone called black or white?

Pamphlets with Issues

Back to the Second World War. Mathematics also served the Allies. Consider the demagogic use of the following mathematical problem that was distributed as a pamphlet.

> *The official German figure for dead and missing on the Eastern Front is 337,342 on 2 July 1942. The number of German families according to the Statistical Yearbook is 21 million. What part of the German families lost a relative on the Eastern Front? Answer: 21 million / 337,342 = 62.25. Thus, the Führer caused 1 in 62 German families to lose a family member on the Eastern Front.*

Shocking too is the true story of the Edelbach concentration camp, now an abandoned village a few kilometres south of the main road from Vienna to Prague. Anna Maria Sigmund, Peter Michor and Karl Sigmund wrote an extensive article about it in The Mathematical Intelligencer. The 5000 predominantly French prisoners of the Offizierslager developed a curious survival strategy by studying mathematics. Their research focused on algebraic topology, under artillery lieutenant Jean Leray (1906–1998). He was the rector of a kind of 'University in Captivity', which awarded some 500°. After the war, many received recognition for their work, as they continued to study at the Sorbonne or the Collège de France (Fig. 2.5).

The prisoners were taught abstract science because the Germans did not see anything suspicious in it. Leray had been a specialist in fluid dynamics and mechanics before the war, but he turned to topology in the concentration camp. Just as the Third Reich looked down on 'Entartete Kunst' (Degenerate Art) as being decadent and morally suspect, they misjudged algebraic topology as an essentially 'useless' branch of mathematics and left Leray alone to focus on it. Yet, the inmates were not afraid of earthly endeavours—literally—they dug a 90-m tunnel through which 132 prisoners escaped in 1943. It was the largest escape from a war camp in World War II.

The story is not well-known. One reason may be that Hitler himself went to great lengths to get Edelbach consigned to oblivion. His father and mother were born in this area, and he had erased all references to his origins as much as possible. But, also born there were Leray's so-called 'spectral rows' and 'sheaf theory', which are now important concepts in topology. Unlike Nazi theories, mathematical ones will live for a thousand years.

Fig. 2.5 The 440 by 530 m of the Oflag VII housed a true 'university in captivity'

Post Scriptum

When this chapter first appeared in Dutch in the science magazine EOS, I dedicated the article to my uncle Omer Huylebrouck. During WWII, he was the leader of the Resistance in East Flanders, was captured and put in prison but surprisingly released during the war. He had escaped a certain death, the family legend said, 'by solving the problem of how many eggs and rabbit legs were needed to bribe the German guards.'

After the publication of the article, I was told that a 'negotiator' had also been involved who 'had good relations with the German occupier'. He had given the bribe as described but was subsequently arrested and died in German captivity. It raised a new question: was not that 'negotiator with good relations with the Germans' who had paid with his life for solving the mathematical problem of eggs and rabbit legs, as great a hero as my uncle...?

3

Guernica

The bombing campaign against the Syrian city of Homs, and the campaigns against Rotterdam, Coventry and perhaps also Dresden belong to the dubious category of terror bombings. However, the first terror bombing in history was that of Guernica, on 26 April 1937.

Fig. 3.1 Picasso's *Guernica* (Mural of the painting *Guernica* by Picasso made in tiles and full size, in Guernica)

Picasso's 1937 Masterpiece

In 1937, Pablo Picasso, then fifty-six, was considered the most important living painter in the world. Thanks to his popularity, he was commissioned by the government of the Spanish Republic to create an enormous mural almost eight meters wide for the International World Fair in Paris in 1937. Picasso, however, had no inspiration for months. But when a Nazi air raid devastated the Spanish Basque city of Guernica (or Gernika in the Basque language), his outrage spawned one of the most impressive works of art of the twentieth century (Fig. 3.1).

Jose Antonio Aguirre, the Chairman of the Autonomous Basque Government during the Spanish Civil War (1936–1939), described the bombing as follows:

> 'Gernika was a town in a peaceful valley, surrounded by mountains, with about seven thousand inhabitants. It was an open city without any means of defence. Gernika held a market day every Monday, a well-known and picturesque event, attracting farmers and villagers. This was the scene that Franco and ally Germany had chosen for the first rehearsal of their total war.'

The city was far away from the front lines, the choice of target was inexplicable, and the attack was simply outrageous. Just days after 26 April 1937, Guernica became a symbol of meaningless violence. Never before had such a carefully planned and at the same time totally unexpected act of terrorism taken place. At the time, George Steer, the special correspondent for The Times in the Basque Country, described the events as follows (Fig. 3.2):

> 'Gernika, the oldest city of the Basques and the centre of their cultural tradition, was completely destroyed yesterday afternoon by the insurgents' air raid. The bombing of this open city, far behind the lines, took exactly three hours and a quarter. All the time, a huge fleet of German fighters and Junkers and Heinkel bombers shelled the town with bombs weighing up to 1,000 pounds and more than 3,000 two-pound aluminium fire projectiles. The fighter planes dived low over the centre of Gernika to use their machine guns and mow down the civilian population who had taken refuge in the fields.'

Fig. 3.2   The destroyed city of Guernica

Picasso on the Warpath

Despite the civil war, the government of the Spanish Republic decided to take part in the international exhibition that was to be opened in Paris in May 1937. It was a strange decision because the country was in turmoil. However, with its participation, the Spanish government wanted to draw attention to the unstoppable rise of fascism in Europe; in other words, it was a propaganda offensive.

The 1937 Spanish Pavilion was designed and built by Joseph Lluis Sert and Luis Lacasa in just four months. From the beginning, the government of the Republic wanted to recruit Picasso for their cause. Sert visited the painter in Paris to convince him to make a large mural. It would become the centre of the building, surrounded by works by artists such as Joan Miró, Julio González or Alexander Calder. The building was to become a milestone in art and architectural history.

Yet, in the months following the assignment of the commission for the Spanish Pavilion, Picasso fell prey to a kind of writer's block. He couldn't get anything on canvas until, on 27 April, all French newspapers reported the bombing of Guernica. The destruction of the Basque city was almost total, and Picasso was so shocked by the tragedy that, from then on, he would focus on a single theme: Guernica. It was not political to him. It was a matter of injustice; he was outraged by the crime against innocent victims. That same day, Picasso began work on the painting that can perhaps be called the most moving and powerful work of art of the twentieth century (Fig. 3.3).

During the first ten days, Picasso was driven by pure anger, as evidenced by the twenty-one sketches and studies he made from that period. On 11 May he cast all his indignation onto the canvas in his studio in the Rue des Grands-Augustins in Paris. Picasso's intentions were clear: 'Painting is not about decorating apartments. It is an instrument of war for attack and defence against the enemy.'

Fig. 3.3 In the footsteps of Picasso's many sketches for *Guernica*. To this day, artists study the work by drawing it. Here is a sketch from 2011 by artist Steffipedia

Picasso continued to paint the huge canvas tirelessly, making twenty-four further preliminary studies and sketches of the main characters. He wouldn't stop until it was completely finished on 5 June. Afterwards, the painting was to be displayed for about two weeks in the newly finished Spanish Pavilion. Picasso's fury and creativity exploded, as his biographer Pierre Daix observes: 'In May 1937, Picasso went to war.'

Geometry and Guernica

The *Guernica* scene is clearly inspired by *The Horrors of War*, a painting by Peter Paul Rubens from 1637. But Pablo Picasso reversed the direction of the action in the Rubens painting (from left to right). The action now moves from right to left, which makes the effect more intense because a viewer's gaze physiologically moves from left to right. This effect is intensified by the chaos of the angular distortions and the violent contrasts of light, shadow and texture (Fig. 3.4).

The mural is built up according to the classical scheme of Greek temples or medieval altarpieces. Two smaller side panels frame a large central panel that is twice the width of the side panels. Within this framework, a large isosceles triangle dominates the composition. The top of the triangle is the top central point, and it connects almost seamlessly with the light rays of a small lamp. The balance achieved by the hands and feet on either side of the triangle is emphasized by the mass of the statue's head, and the large knee of the walking woman. The composition has been carefully thought through (Fig. 3.5).

Fig. 3.4 Same theme, different era and painter: *The Horror of the War* (or *The Consequences of War*), by Peter Paul Rubens (1637)

Whether the action takes place indoors or outdoors is not clear. In the first sketches, the sun clearly dominates the scene, and several buildings are clearly visible. In later studies, the sun changes into an oval shape and eventually Picasso adds a light source. At the same time, windows and doors replace the buildings. And so, the viewer is guessing whether it is an indoor or outdoor scene. The fear and confusion are manifest: for the inhabitants of Guernica there is no way to escape the horror.

Endless Tinkering

While Picasso was painting, his then partner Dora Maar took a series of ten photographs. Together with Picasso's 45 preliminary studies, these show that he kept working on the canvas until the very end. The horse and the bull, the most important elements of the composition, underwent numerous transformations. From a geometric point of view, the bull's dominant position changed by turning the body to the left; however, the bull's head did not move. The bull's flame-like tail took the place of the moon that was on the left side of the canvas in earlier drafts. The shift of the bull's body away from the centre also served to make room for the horse's head and for a bird, probably a pigeon.

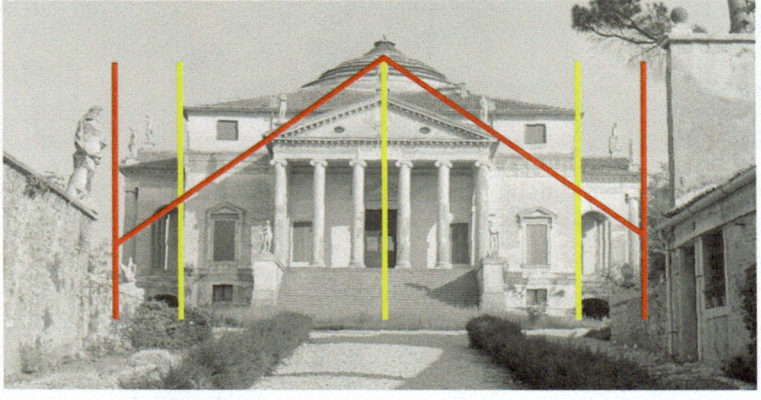

Fig. 3.5 *Guernica*'s geometric construction can be compared to the design of Renaissance buildings such as Andrea Palladio's Villa La Rotonda (1566)

The place previously occupied by the back of the bull became the location of the horse's head, drawn in a grimace of pain. The bull, the only stable and undistorted figure in the composition, seems to be the only hope of rescue to all other characters. It turns its head straight towards the viewer with a noble and humane face; that of the painter himself?

The warrior's body was the last to undergo changes. On 4 June—barely a day before the painting's completion—Picasso made his last two sketches: a study of the warrior's hand and head. His dead body was now torn in two, emphasizing the absolute destruction visited upon the town in the bombing. The head is ovoid, almost without details, thus referring even more clearly to death. A flower emerging from the broken sword can be seen as a sign of hope and rebirth. Picasso's *Guernica* was finally finished.

To This Day

This text was based on a lecture given by Javier Barrallo, a computer scientist from the University of San Sebastian—Donostia, Spain—Basque Country. He collaborated with the computer graphic artist Lena Gieseke, from the University of Georgia (USA), to create a 3D interpretation of the painting. Their intention was to revive the attention to the forgotten Spanish civil war. But under pressure from the Picasso Foundation, which invoked copyright rules, Gieseke took her work off the Internet and cancelled all her presentations. Barrallo however still shows the film, because, according to him, 'it is unacceptable that foundations and film directors get all the money, and the Basques the bombs.'

Barrallo also disagrees with the title of '33 days', the film about Picasso's *Guernica* by Spanish film director Carlos Saura. Antonio Banderas played the role of Picasso, though he had to use a lot of make-up. However, he himself was very keen on it because he was born in Malaga, like Picasso, and because he walked past his house every morning on his way to school. According to Picasso's daughter Paloma, Banderas has the same accent as her father. The title '33 Days' would refer to the number of days it took to paint *Guernica*, but this is a mistake, Barrallo argues. A minimum of 35 and nearly at least 36 days passed between Picasso's first sketch on 1 May 1937, and the last on 4 June 1937. That is 35 days, and it took Picasso an additional day to finish some details.

To this day, there still are witnesses who saw Picasso's *Guernica* in the original scenery. At the age of fifteen, Belgian artist Octave Landuyt visited the Spanish pavilion in the Expo in Paris in 1937, where Picasso's *Guernica* was exhibited. In 1953 he was introduced to Picasso by Emile Langui who was then working for the 'Art Propaganda Service' at the Ministry of National Education. At the time, Langui had clandestinely supported the actions of the government forces against Franco. Another common acquaintance and comrade in arms of Picasso and Landuyt was writer and politician André Malraux, who had participated in the Spanish civil war with the financial support of Picasso. In turn, Landuyt illustrated poems by the Spanish poet and playwright Garcia Lorca, who was murdered by Franco's followers. Landuyt is a first-class witness: 'Picasso made a preliminary study for the life-size sketches of *Guernica* using cut-out pieces of newspaper. This is an ideal method to test a creation of that size' (Fig. 3.6)

Fig. 3.6 Left Picasso, right Belgian painter Octave Landuyt, drawn by Willy Bosschem

4

Architect-Alchemist

Some French authors have already torn down Le Corbusier from his pedestal by exposing his fascist past. Now his most basic mathematics turns out to be wrong too. In fact, his famous 'Modulor' was deliberately misleading.

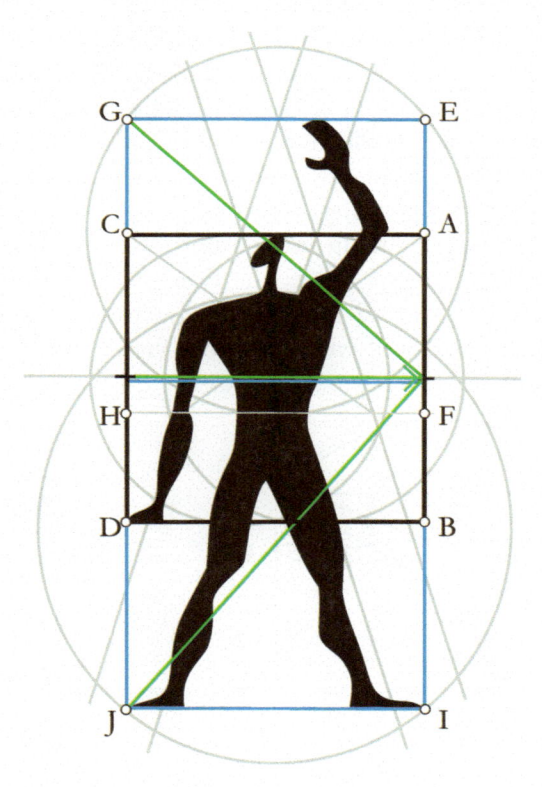

Fig. 4.1 Like a medieval alchemist, Le Corbusier drew mysterious arcs and triangles, without understanding what they represented

Racist Proportions

The 'greatest architect of the twentieth century' is undoubtedly Le Corbusier (1887–1965). That name was the pseudonym of Charles-Édouard Jeanneret-Gris, a Frenchman born in Switzerland. The literature is stuffed with exhilarating statements about his work. For example, the equally French architect Jean-Louis Cohen (University of New York) stated that Le Corbusier's so-called 'Modulor' was probably 'the most comprehensive proportional system of the twentieth century.'

However, a recent publication by mathematician Natasha Rozhkovskaya (Kansas State University, USA) shows that the Modulor was utterly wrong from a mathematical point of view. In fact, it was a deliberate deception (Fig. 4.1).

It is well known that Le Corbusier was a great proponent of the golden ratio, the ratio that divides a line segment according to the number $\phi = (1 + \sqrt{5})/2 = 1.618\ldots$. In 1854 German psychologist Adolf Zeising created the myth that this is the measure of ideal beauty. The 'most beautiful rectangle', say 'miss USA rectangle', would have a length of 1.618… and width of 1. The Greeks would have used this golden rectangle in the Parthenon, and Leonardo da Vinci in his paintings (Fig. 4.2).

All these stories were made up in the mind of the creative German, reinforced by another mysterious figure, the Romanian 'prince' Matila Ghyka. Their ideas got widely spread and eventually reached Le Corbusier, who perhaps thought it fitted well with his fascist views. For example, a person

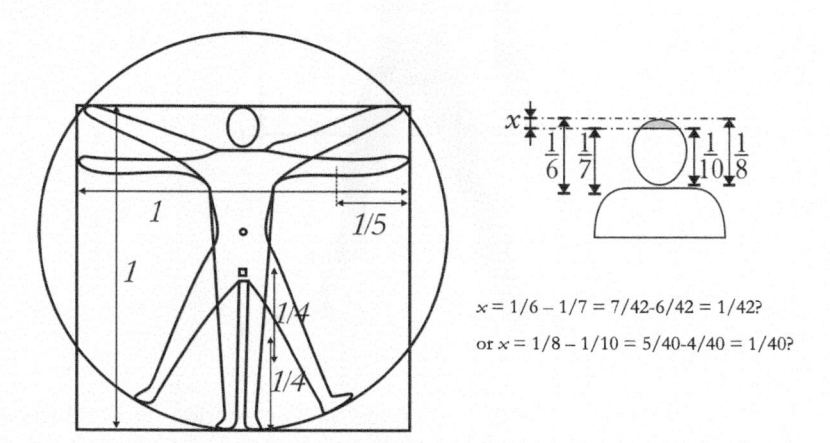

$x = 1/6 - 1/7 = 7/42 - 6/42 = 1/42?$
or $x = 1/8 - 1/10 = 5/40 - 4/40 = 1/40?$

Fig. 4.2 Leonardo's Vitruvian man has nothing to do with the golden ratio and describes with simple fractions the proportions in man, such as from the neck or the chest to the forehead and to the top of the head. Sometimes, they clearly contradict each other

of 1.62 m must have the navel at a height of 1 m, otherwise he or she is an *Untermensch*. Books promoting these theories state that these relationships hold for Westerners, but not for African or Jewish people (Fig. 4.3).

That, of course, suited Le Corbusier well. In *Le Corbusier, un fascisme français* (Le Corbusier, a French fascism, 2015) journalist Xavier de Jarcy showed that Le Corbusier was politically incorrect. He was joined by authors such as François Chaslin and Marc Perelman. In a letter to his mother, for example, Le Corbusier expressed his admiration for fascist ideas:

> 'Money, the Jews (who are partly responsible), the Freemasons, all will feel law and order – Hitler can crown his life with a great work: the planned layout of Europe.'

However, many people closed their eyes and let the storm blow over. They stated that the architect cleared his name through his oeuvre. They compared him, for example, to rocket designer Wernher von Braun, who also had Nazi sympathies but sent the first American to the moon as leader of the Apollo program. It's a curious argument. Why, for example, does it not apply to collaborators who were good plumbers or punctual tram drivers?

In fact, much of Le Corbusier's work bears witness to his fascist ideas: he was a strong proponent of standardization in just about everything—from material to structure and to design. This resulted in those characteristically

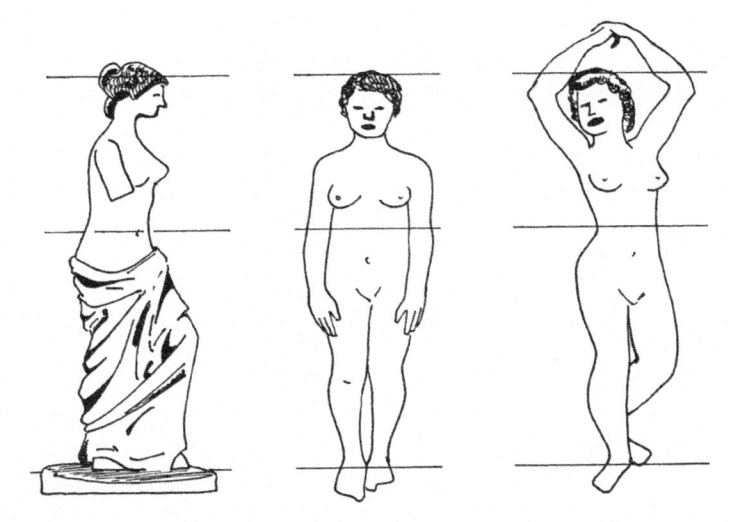

Fig. 4.3 The navel is perfectly positioned at 1 m for a Greek woman of 1.618 m (left), but not for an African and Jewish woman (centre, right; drawing D. Néroman, *Le Nombre d'or, clef du monde vivant*, Ed. Dervy, 1981)

Fig. 4.4 The Cité Radiant, or Radiant City in Rezé, France, radiates less than the church in front of it

soulless concrete pigeon boxes that some euphemistically classify as 'brutalist architecture'. French philosopher Roger-Pol Droit described such conformed housing units as 'concrete prisons designed to format people. They stand very far from ideas of freedom and human rights. And very close to Mussolini's dream' (Fig. 4.4).

Modulor

Le Corbusier invented a lengthy pseudo-scientific bogus story about his golden ratio theory, his ratio system called the 'Modulor'. It consists of two lists of numbers. One list contains approximate doubles of the numbers in the other list. Each number in a list follows from its preceding term by multiplying it by the golden ratio $\phi = 1.618\ldots$ and rounding it off—or not.

Le Corbusier wanted to gain a reputation as an expert in standardization. That is why he announced in 1948 that he had a solution for a 'harmonious standardization of mass production based on mathematical foundations.'

4 Architect-Alchemist

Lacking modesty, he suggested using his 'invention', the Modulor, as the one and only reference in the design of new buildings and in urban planning.

And he succeeded: even today it has a prominent place in the history of art and architecture. It features in academic courses for architects and designers and inspires 'intellectuals' who see it as a 'symbiosis of mathematics and art'. Though this is mainly due to their lack of knowledge of mathematics. After all, the abundant calculations and geometric doodles in Le Corbusier's writings do not go beyond the level of the first year of a good high school.

Le Corbusier obscured the simple matter of proposing a series based on ϕ by dividing his Modulor into 'red and blue' numbers and placing a man with a raised hand next to it. The numbers would represent his body proportions, but that was done randomly—the drawing was adjusted according to the numbers. The man was at first 175 cm, later he became 182.9 or 183 cm. When it suited the architect, he rounded off some of the numbers in the list (those numbers are in brackets on the drawing).

Thus, the blue numbers would be the doubles of the red. That's very approximate though: $33 = 2 \times 16.5$ but $86.3 \approx 2 \times 43.2$.... Le Corbusier was also proud that his numbers resulted from the addition of each other: $4.8 + 7.8 = 12.6$; $10.2 + 16.5 = 26.7$; ... but that too was 'about' as good as it got: $33 + 53.4 \neq 86.3$ and $69.8 + 113 \neq 182.9$. Moreover, that additive property is not so remarkable. Take a geometric sequence, such as a, ax, ax^2, ax^3 ... in which the base number a is always multiplied by x. Now, require that this sequence can also be realized by addition: that $ax^2 = a + ax$, or $x^2 = 1 + x$. This implies $x = (1 \pm \sqrt{5})/2 = \pm 1.618...$, and $+1.618...$ is indeed the golden ratio. That is not rocket science.

Le Corbusier arrived at his Modulor via a two-dimensional grid with right angles and all kinds of theories about what a harmonic composition should be. His assistant Jerzy Soltan rebuked him:

> 'It seems to me that your invention is not based on a two-dimensional phenomenon, but rather on a linear phenomenon. Your grid is just part of a linear system, a series of golden section divisions that go to zero on one hand and infinity on the other.'

Indeed, Le Corbusier had described something trivial: two simple geometric sequences with factor ϕ.

'Mathematics' by Le Corbusier

Le Corbusier had thought that his Modulor construction was two-dimensional, because in 1943 he had searched for the solution to a problem that he had posed to another assistant, Gerald Hanning:

> 'Take a man with a raised arm, 2.2 meter tall, place him in two squares of 1.1 feet by 1.1 feet each, one above the other, and put a third square astraddle above these first two squares. This third square should give you the solution. The location of the right angle should help you decide where to place this third square.'

The question is vague, but the intention becomes clear when one looks at the images that go with it. Le Corbusier thought the solution had something to do with the golden ratio (Fig. 4.5).

He said that one had to draw the line AM first, from the centre M of the side CD of the square ABCD with side 1. Then according to the Pythagorean theorem $AM^2 = 1^2 + \frac{1}{2}^2$ and therefore $AM = \sqrt{5}/2$ (Fig. 4.6).

Then he turned AM around M so that GM lays on the line CD (Fig. 4.7).

Now if N is the middle of AB, then again according to the Pythagorean theorem $GN^2 = GM^2 + MN^2 = 5/4 + 1$ and so $GN = 3/2$ (Fig. 4.8).

Now consider GN as the right side of a right triangle GNJ with right angle in N. Then, the triangles NGJ and MNG are similar. And so, if $x = GJ$, then $x/GN = GN/GM$ and so $x = 9/(2\sqrt{5}) = 2.01246...$ (Fig. 4.9).

If we now place a rectangle with height 1 on GJ, then it looks as if the square ABCD has been doubled to the rectangle GJEI (Fig. 4.10).

However, that is only the case on a roughly executed drawing, because GJ = 2.01246... and not 2, although it seems so, approximately.

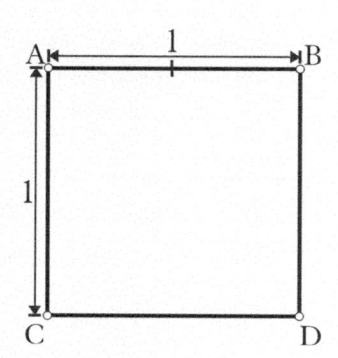

Fig. 4.5 Consider a square ABCD with side 1

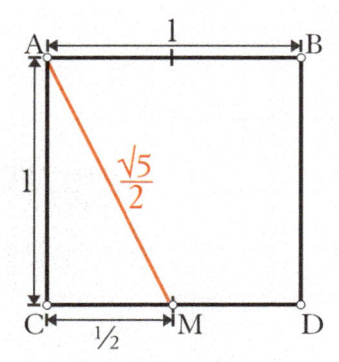

Fig. 4.6 The length of the segment from a vertex to the centre of an opposite side is $\sqrt{5}/2$

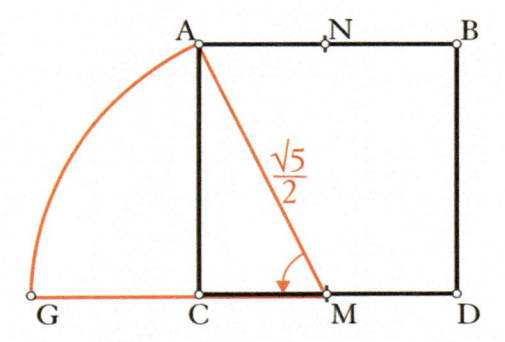

Fig. 4.7 Turn the line segment around that centre; of course, the length of GD then equals $\sqrt{5}/2 + 1/2$, the golden section

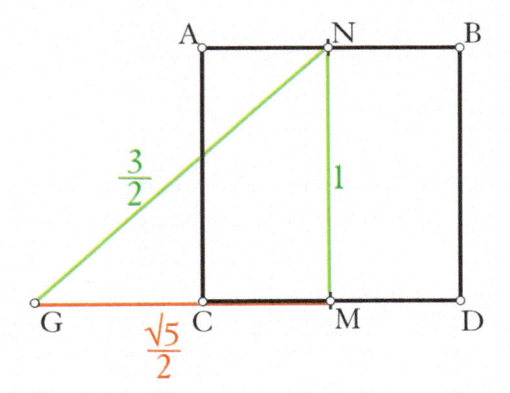

Fig. 4.8 The hypotenuse in the right triangle GMN measures 3/2

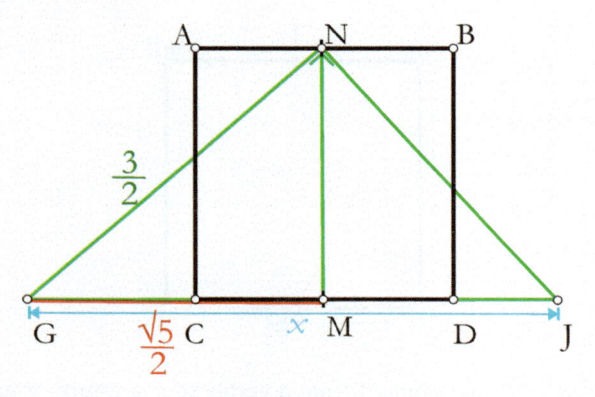

Fig. 4.9 The Pythagorean theorem teaches that $x = 9/(2\sqrt{5})$

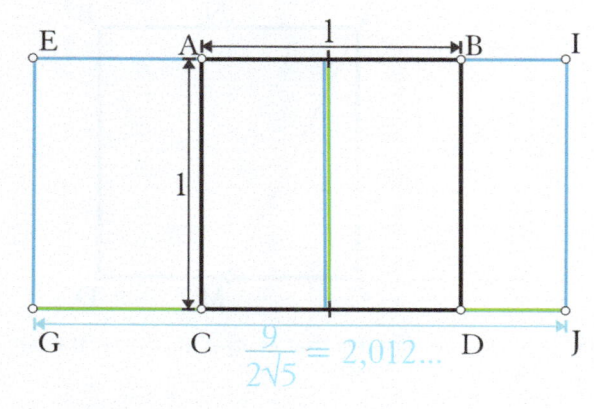

Fig. 4.10 On a roughly drawn sketch, it looks as if GJEI is 'twice' the square ABCD

Deliberately Wrong

Le Corbusier shared this idea widely in his writings. It regularly returns in his work, as in the opening illustration of this chapter. He considered his solution for the doubling of a square as a great mathematical discovery. That's bad luck because it is simply wrong. It resulted from Le Corbusier's inaccurate drawings, without relying on logical reasoning and actual mathematics.

Moreover, the flaw fitted well with one of the few mathematical constructions known to Le Corbusier, that is the drawing of the golden ratio based on a square. After all, in the second drawing of the foregoing construction, the length of GD is equal to that of GM plus MD. That is $\sqrt{5}/2 + 1/2 = (\sqrt{5} + 1)/2$, which is the golden ratio. However, doubling a square by drawing (i.e., with a ruler and compass in a finite number of steps) with a right angle

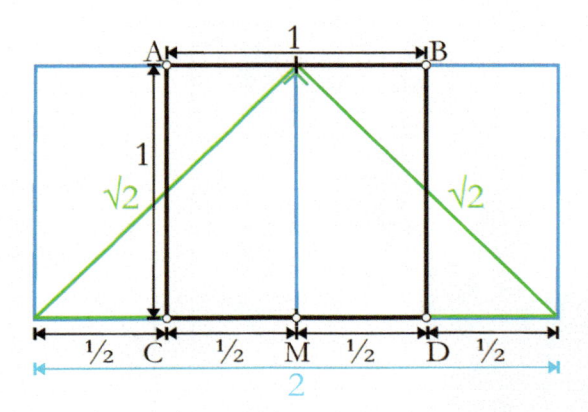

Fig. 4.11 The doubling of the square ABCD by a right angle can be done very easily by placing it in the middle

is simple: place a right triangle in the middle, the rectangular sides of which have the length $\sqrt{2}$ (Fig. 4.11).

Le Corbusier's doubling of a square with side 1 into a rectangle with width 1 and length 2.01246… has a deviation of 0.6%. Some may argue that this error of only 0.6% is insignificant, but that would contradict Le Corbusier's concepts of exact harmony and ideal composition. After all, perfectionism was the goal he pursued. Moreover, he absolutely wanted his invention to be of a mathematical-scientific nature.

In addition, Le Corbusier proposed that his solution would be used in standardized dimensions. However, a systematic error of 0.6% poses significant potential problems for engineers and technicians (Fig. 4.12).

In addition, Le Corbusier was aware of the absurdity of his mathematical manipulations. In 1948, he asked the mathematician René Taton for advice. Taton replied, of course, that the doubled 'square' was actually a rectangle. The architect laconically dismissed this response. Malcolm Millais, a critic of Le Corbusier, derided his approach, observing that it was the result of a lack of mathematical education:

> 'Just as a medieval alchemist wanted to find a way to turn an ordinary metal into gold, Le Corbusier scratched in geometrical riddles to deliver two [geometric] rows. […] Le Corbusier saw this as an enormous achievement'.

However, Le Corbusier knew very well he was wrong. He persisted because his Modulor was a marketing stunt for self-promotion. And a successful one: architects and artists have been falling for it for more than fifty years now.

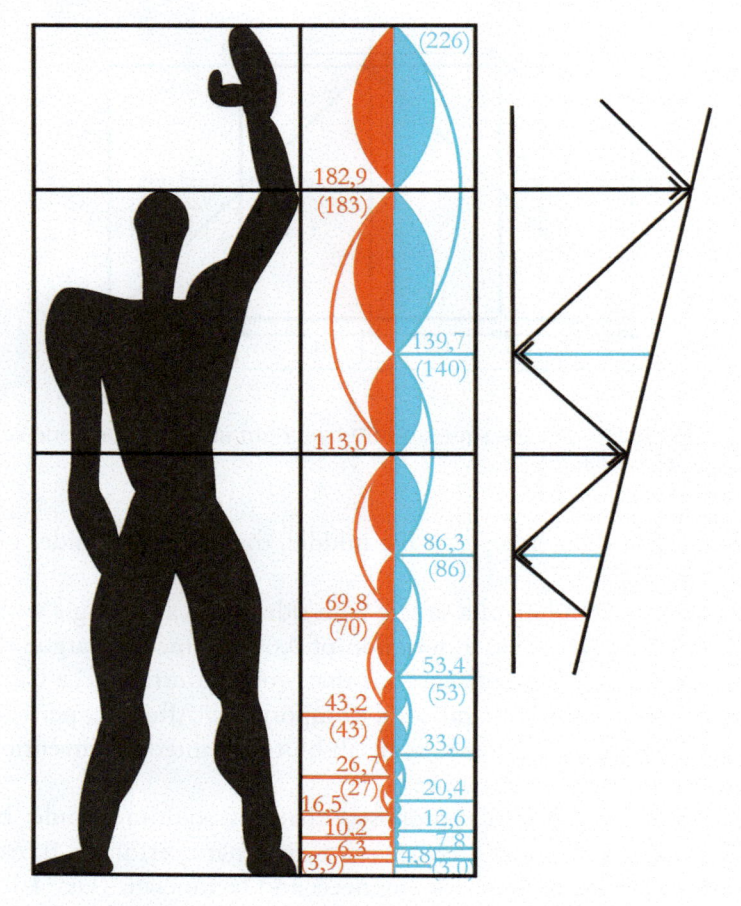

Fig. 4.12 The Modulor, with the 'principle of the right angle'

Just as an alchemist mixed frog legs with sulfuric acid to turn metal into gold, Le Corbusier drew circles and lines to double the square ABCD, so to speak, into the rectangle EIJG.

Golden Ratio Buildings

We should not throw the baby out with the bath water. Although the golden ratio is not a necessary condition for something to be beautiful, this does not imply that a design with the golden ratio cannot be beautiful. Le Corbusier's Chappelle Notre Dame du Haut in Ronchamp is considered a masterpiece, but whether this is due to the golden ratio remains to be seen. Besides,

4 Architect-Alchemist

Fig. 4.13 Chinese copy of the Chappelle Notre Dame du Haut on which a golden section rectangle was drawn (the French panorama law prohibits the publication of photos of the original building)

anyone who wants to see the golden ratio in it must show a lot of good will (Fig. 4.13).

Le Corbusier had so many followers, one can discover golden section buildings everywhere. For instance, there is a beautiful post office building, now serving as a cultural centre, in the author's hometown of Ostend, Belgium. The designer, Belgian architect Gaston Eysselinck, was indeed a follower of Le Corbusier, like so many others. His building was designed to fit neatly in a golden rectangle. Eysselinck, however, was not a dogmatic believer in the golden section and was a very creative architect in his own right. He placed a majestic work of art on the building, despite great opposition from the town officials. Did Eysselinck feel he had to improve Le Corbusier's pseudo-scientific architectural approach? (Fig. 4.14).

Fig. 4.14 The Ostend post office fits into a golden rectangle

5

War Hero, Math Genius, Martyr

Mathematics has a martyr too: mastermind Alan Turing cracked Hitler's code but died in suspicious circumstances from the poison of a cyanide apple.

Fig. 5.1 Sculpture of Alan Turing in Bletchley Park (England). It is made of slate from North Wales, where Turing used to spend his holidays

Ultimate Recognition for Gay Icon and Top Scientist Alan Turing

According to Time magazine, Alan Turing was one of the twenty most important scientists and scholars of the last century. The mathematician cracked a vital code of the Wehrmacht during the Second World War, and the consequences were so decisive that films, plays and cartoons were made about it. To many computer scientists, Turing is also the father of modern computer science, while his achievements in the most abstract mathematics are studied to this day. He died in 1954, just a few days before his 42nd birthday, from a dose of cyanide.

The circumstances of his death remained mysterious for years: a partially eaten apple was found next to his bed. There is a strong suspicion that the apple had something to do with his death, but no one ever knew exactly what. Or perhaps no one wanted to know, as Andrew Hodges concluded in his biography *Alan Turing, the Enigma* (1983). At that time there seemed to be no memorial in honour of Turing, a war hero, and a hero of science.

However, this changed, step by step. In 2001, the University of Manchester placed a lifelike sculpture of Turing in the small Sackville Park in the centre of the city. The traditional media reported little about it, although the specialized mathematical press and LGBT media did make an effort to motivate visitors to Manchester to make a pilgrimage to the statue. And so, at the beginning of the twenty-first century, Turing's name was not yet cleared, but he did gradually get the recognition he deserves (Figs. 5.1 and 5.2).

Life and Work

Alan Mathison Turing was born in London on 23 June 23 1912. He attended Sherborne College in Southwest England. At the time, mathematics and the sciences were looked down upon, as popular opinion held a classical education in much higher esteem. As a result, he ended up at King's College, Cambridge, and not at the more prestigious Trinity College. After his studies, he received a scholarship as a researcher, based on his dissertation on the central limit theorem. Many readers will remember this theorem from their statistics courses: the mean of a large number of independent variables approximately follows a normal distribution (a bell-shaped Gaussian curve), even in the case that the variables themselves do not follow a normal distribution (Fig. 5.3).

Fig. 5.2 Turing's dramatic life inspired many stories, such as this cartoon about mathematics and military espionage. The title refers to Turing's childhood friend

Fig. 5.3 16-year-old Alan Turing

At the age of twenty-four, he published the article 'On Computable Numbers', which led many to consider him the father of computer science. The subtitle was 'with an application to the entscheidungsproblem'. This refers to a question by David Hilbert about decision theory: is there an algorithm (a finite set of clearly formulated statements), drawn up in a formal language so that a mathematical statement in that language can always be clearly classified as true or false? Of course, one would think that is what mathematics is all about! Actually, it isn't: between 1935 and 1937 Alonzo Church and Alan Turing formulated clear, but negative answers. They were based on the work of the famous logician Kurt Gödel, by reformulating the abstract approach of his logic to the level of computer actions.

This resulted in the 'Turing machines': not real machines with electronics or cogs, but very accessible and intuitive thinking constructs. Which gave birth to an entire field concerned with the 'theory of computation', as it is called today. The idea of a universal Turing machine that could take over all thinking tasks from other machines created the hope for a computer maid-of-all-work, or more respectfully, a machine-of-all-thinking.

At the time, Turing worked with Church at the prestigious Princeton Institute for Advanced Study, where he defended his doctorate in 1938. He then returned to Cambridge. War soon broke out, and his interest in coding theory led Turing to the *Bletchley Park Code and Cypher School*. He was recruited by the government and improved the Polish Bomba decoding machine, which was the only one capable of cracking the German Enigma code at the beginning of the war. Turing's improvement turned out to be a decisive breakthrough in WWII. Undoubtedly, deciphering the messages from the German naval forces saved hundreds of thousands of lives. In 1945, Turing was awarded the Order of the British Empire for his services during the war (Fig. 5.4).

Fig. 5.4 The Provincial Domain of Raversyde, Belgium, displays an Enigma coding machine in one of its bunkers

After the war, Turing joined a research group at the University of Manchester responsible for many breakthroughs in the development of the electronic calculator and the computer. One of Turing's most important contributions was a surprising and very readable manual on computer programming, so that he can rightfully be considered the very first 'computer whiz'. At that time, in 1948, Turing even played the role of a computer to test one of the first chess programs, as no real computer was yet powerful enough to test the program (admittedly, Turing lost).

In 1950, Turing described an artificial intelligence experiment that still bears his name: The Turing Test. It was ahead of its time, and it is therefore not surprising that the Turing Award, given annually by the Association for Computing Machinery since 1966, is widely regarded as the ICT Nobel Prize.

The Turing Test

The thought test that still bears the name Turing was a challenging contribution to the field of artificial intelligence in the 1950s. It is one of the first tests to have been proposed to determine whether a machine could demonstrate intelligence. Turing described it in his paper *Computing Machinery and Intelligence*; a group of (human) jury members communicate with a machine and with a human, isolated from each other. The machine, which tries to display human behaviour, passes the test if the jury cannot distinguish it from the human.

While some television shows may give the impression that the comparison with a human does not necessarily imply a proof of intelligence, the Turing Test was a particularly bold statement on the intelligence of computers at a time when there were hardly any. Turing suggested that it would be a good move for the machine to mimic some human errors and thus display unintelligent behaviour, on purpose. Moreover, the composition of the jury was another weak point because which people can be considered smart enough to determine whether the highest intelligence in the world is a human or a machine?

The Turing Test is more noteworthy than ever, as many modern customer service systems use chat bots to field queries. Until its final competition in 2019, the Loebner Prize featured a similar test in which a machine had to convince at least 30% of the interrogators that it was human. The jury consisted of philosophers, computer scientists and journalists. In the autumn of 2008, the chat robot Elbot of the Artificial Solutions

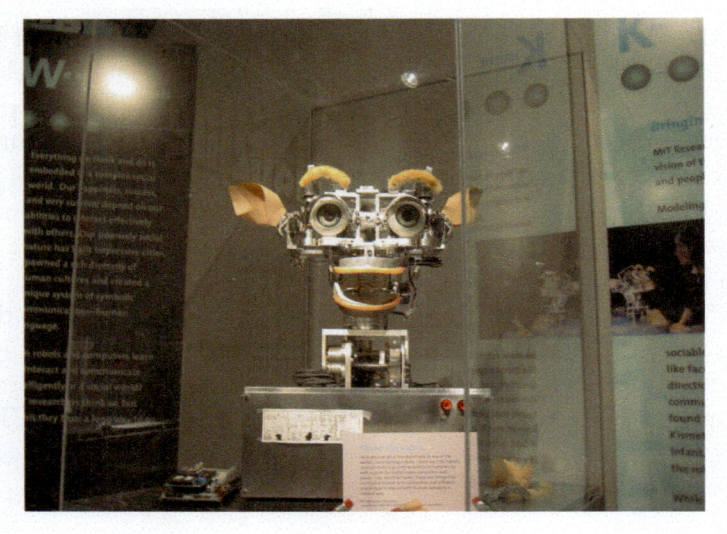

Fig. 5.5 The KISMET robot in the MIT museum also attempts to convey human emotions through facial expressions

company managed to make 25% of its conversation partners believe that they were conversing with a human. Initially, conversations only involved text exchanges via keyboard, but today there are also test variants with speaking virtual characters (Fig. 5.5).

Mathematical Nuance

Although Turing's contribution to computer science can hardly be overestimated, and he is sometimes described as some sort of Bill Gates avant-la-lettre, he was a true mathematician. Not only was his master's thesis concerned with statistics, but later on Turing also became interested in many other areas of mathematics. Towards the end of his life, he focused on the biological processes by which an organism develops certain shapes and not others (such as determining the number of leaves of a plant, which follows the well-known Fibonacci sequence: 1, 1, 2, 3, 5, 8, …).

For a long time, Turing also focused his attention on the so-called zeta function $\zeta(s)$, noted with the Greek letter zeta or ζ. It was coined by the German mathematician Bernhard Riemann. His 'Riemann Hypothesis' was made famous by the film about John Nash, *A Beautiful Mind*. For the values 2 and 3 it is given by:

$$\zeta(2) = 1 + 1/2^2 + 1/3^2 + 1/4^2 + \cdots$$
$$= 1 + 1/4 + 1/9 + 1/16 + \ldots = 1.6\ldots$$
$$\text{and } \zeta(3) = 1 + 1/2^3 + 1/3^3 + 1/4^3 + \ldots$$
$$= 1 + 1/8 + 1/27 + 1/64 + \ldots = 1.2\ldots$$

It is also possible to use exponents in the denominators that use one of the hypothetical roots of -1 (with mathematical symbol i), in which case this function can become zero. Riemann wondered for what values this happens and made his very surprising suggestion, which computer calculations would always confirm later on, but could not prove.

For one reason or another, the Riemann hypothesis would fascinate many important mathematicians. Turing's work on it is not necessarily his most famous, but it is one of the areas where the power of his mathematical thinking is felt most profoundly. In 1950 he made one of the first attempts to calculate the sought-after zeros of the zeta function on a computer; he used the Mark 1 of the University of Manchester, one of the very first computers ever, though today it is only relevant for the archaeology of computer science. In 1939 he had already tried to build a calculator with cogs and all the trimmings, specifically for calculating values of $\zeta(s)$.

A few years later, his abstract mathematical insight and technical skills would help to end WWII. It illustrates how fundamental research in an abstract mathematic field can have unexpected consequences, although films about breaking a code in WWII usually do not focus on that aspect.

A Tragic Death

While it is understandable that screenwriters do not necessarily have a passion for mathematics, it is a pity that movies sometimes deliberately cover up more personal matters. The 2001 film *Enigma* for example, with Dougray Scott and Kate Winslet in the lead roles, found it necessary to disguise the fact that the leader of the team of code breakers was gay. That character was without a doubt based on Alan Turing, who was always open about his sexual orientation, and so there was no reason to suppress this fundamental detail.

In his teens in Sherborne, Turing bumped into an older college student Christopher Morcom, who became his first love. He died of bovine tuberculosis from drinking contaminated milk, a tragic start to Turing's love life. It would also end tragically. In 1952, Turing went to the police to report a burglary, to which his 19-year-old lover Arnold Murray was an accomplice.

During the investigation, Turing admitted having a relationship with Murray, after which the investigator decided to prosecute the victim as well as the perpetrator for illegal sexual acts. Turing had to choose between a prison sentence and hormone injections to cure his 'perversion'. He opted for the injections. The event, however, did have the consequence that Turing could no longer continue his life's work, because for reasons of national security he was no longer allowed to work on code theory. Two years later, in 1954, he was found dead in his home after apparently having taken a bite from an apple poisoned with cyanide. Strangely enough, the apple was never tested by the police, although the autopsy did reveal the presence of cyanide. Was it an accident caused by Turing's habit of handling dangerous chemicals rather carelessly? Or was it suicide—inspired by his favourite story *Snow White*— and did he want to spare his mother's pain by disguising it as an experiment gone wrong? Or were the security forces, who considered a homosexual code breaker a danger, responsible?

Repentance Comes After Sin

Andrew Hodges' 1983 biography inspired dramatist Hugh Whitemore to write his 1986 play *Breaking the Code*. During numerous performances in London and New York, Derek Jacobi interpreted the role of Alan Turing in such a convincing way that a reworked version was filmed for television in 1997. The city of Manchester too began to realize that Turing was a unique figure, who deserved much greater fame than he had to that point. Today, there is an Alan Turing Bridge, a road, and several buildings bearing the name of the genius mathematician. At the University of Surrey, the area where Turing grew up, a bronze statue of Turing was unveiled in 2004, and three years later, a slate stone statue embellishes Bletchley Park, where Turing cracked his codes.

The Alan Turing Memorial in Manchester's Sackville Gardens, unveiled in 2001, has a curious history. The initiative came in 1997 from some enthusiasts who were already supported by the British Society for the History of Mathematics, though the computer industry remained remarkably aloof. Fortunately, there were other financial contributions, from the City Council and the gay community in Manchester. Also noteworthy were the many small donations from older people, who wanted to show their gratitude to a hero from WWII, and so the statue would be unveiled four years later. It showed a life-size likeness of Turing, next to whom one could sit down, and not a classic upright image of a stern figure looking to the heavens. It worked, because

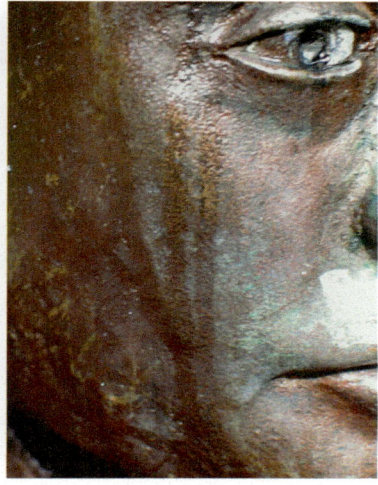

Fig. 5.6 The statue in Sackville Gardens, which shows peculiar lines under one eye, as if it sheds a tear

there were sometimes lines of people waiting their turn to sit next to the statue of Turing. Sculptor Glyn Hughes symbolically buried his old Amstrad computer under the memorial plaque by the statue. He also incorporated bronze from guns in the statue, and the different metals caused peculiar lines under one eye of the statue, as if it sheds a tear (Fig. 5.6).

Other Turing enthusiasts include Pieter Geelen, co-founder of the navigation system TomTom, who sponsors a Turing Foundation. Yet, the story that the Apple logo refers to Turing's poison apple (with one big bite) seems to be a fantasy, though it does illustrate how Turing appeals to the imagination.

In 2009, the then Prime Minister of Great Britain, Gordon Brown, decided, after some apparently very persuasive petitions, that he would give Turing an official rehabilitation. Yet, when the next government took office, led by David Cameron, the decision was reversed. Fortunately, Queen Elizabeth II stepped in, giving Turing a rare 'royal pardon' in 2013. And in 2014, a Hollywood-style movie, *The Imitation Game*, also based on Andrew Hodges' biography, was released. Lead actor Benedict Cumberbatch made Alan Turing known to the general public via his powerful and sympathetic portrayal.

What will perhaps secure a greater legacy, however, is the Bank of England's decision to portray Alan Turing on the £50 note, as of 23 June 2021, his birthday (Fig. 5.7).

Fig. 5.7 The new £50 note, with Alan Turing on the reverse (photo courtesy of the Bank of England)

6

Murder and Higher Math

In 2020, American murder convict Christopher Havens announced a top mathematical publication to the world, from his cell. Multiple murderer André Bloch previously wrote an entire mathematical oeuvre from a Parisian psychiatric institution. In Belgium, David Boeckx, imprisoned for 21 years for murder, may not reach the same level, but he does want to study mathematics online at the author's KU Leuven University, be it from behind the prison bars of the Leuven-Central penitentiary.

Fig. 6.1 Behind the walls of the Leuven-Central prison, Belgium, someone is studying mathematics

© The Author(s), under exclusive license to Springer Nature
Switzerland AG 2023
D. Huylebrouck, *Dark and Bright Mathematics*, Copernicus Books,
https://doi.org/10.1007/978-3-031-36255-2_6

Prisoners Find Their Passion in Numbers and Formulas

On 18 June 2020, Boeckx's lawyer Jos Vander Velpen introduced a special question to the civil court in Leuven, Belgium: six convicts wanted permission to follow online lessons at KU Leuven University. Three of them wanted to study computer science, another history, another natural science. And then there is David Boeckx. He had set his mind to math.

The man had already obtained several certificates showing that he had passed tests in the exact sciences, so his application cannot be dismissed as a momentary fad, said his lawyer Vander Velpen. The KU Leuven University allowed him to study, but in the current educational system, access to the university's digital education platform had become indispensable.

That is the rub. Boeckx was allowed to study by the Ministry of Justice, but due to the prevailing prison rules, access to the internet was extremely limited. For his part, Vander Velpen believed that it was technically possible to give detainees access to that specific educational platform only. Hence the lawsuit.

But why on earth study math, and not law, for example, to defend himself better in court? Or music or visual arts, to be able to be creative in the cell? In the life of the fifty-year-old Boeckx, there were no indications of a predilection for math: he obtained his high school diploma and then immediately started working in the construction industry. That is, until he murdered a prostitute in Antwerp in 2005, and repeatedly raped his ex-wife, so that he ended up in jail for 21 years.

Meanwhile, he had already served sixteen years of his sentence (Fig. 6.1). The remaining five years were just enough to earn a university degree, Boeckx thought. At first, he wanted to study natural and environmental sciences, but the prison management did not like the idea of chemistry tests in a laboratory. Mathematical formulas explode less but do have practical applications too. Boeckx understood that: with a mathematics degree, he hoped to find work more easily upon release. To the surprise of many, perhaps, his main motivation for studying mathematics was: 'It gives me perspective.' In 2021, Boeckx and his fellow detainees won their case: they were allowed to access the online learning platform of the KU Leuven University. Hopefully this was an additional motivation for them to continue on the right scientific track.

Food, Math, Sleep

In the United States, the 40-year-old Christopher Havens was a few steps further ahead. In January 2020, together with Stefano Barbero, Umberto Cerruti and Nadir Murru of the University of Turin, Italy, he published an article from his cell in the mathematics journal *Research in Number Theory*.

In Seattle, Havens was sentenced to 25 years for murder, of which, by 2020, he had served nine years. He had dropped out of high school, couldn't find a job, and became a drug addict. But his life changed in prison. During a months-long solitary confinement for misconduct, he found his passion: mathematics.

In 'the hole', Havens decided to follow the 'Intensive Transition Program', a one-year project in which he followed a strict schedule: 'Eating, doing math, thinking, cleaning, washing.' He taught himself higher math, although the books he ordered by mail were often blocked by the prison management. At one point, handbooks were no longer sufficient, and he wrote a letter to the Mathematical Sciences Publishers:

'To whom it may concern, I'm interested in finding more information on a subscription to Annals of Mathematics for personal use. I'm currently serving 25 years in the Washington Department of Correction, and I've decided to use this time for self-betterment. I'm studying calculus and number theory, as numbers have become my mission. Can you please send me any information on your mathematical journal? Christopher Havens, #349034

PS. I am self-teaching myself and often get hung on problems for long periods of time. Is there anyone who I could correspond with, provided I send self-addressed stamped envelopes? There are no teachers here who can help me so I often spend hundreds on books that may or may not contain the help I need. Thank you.'

The production editor's partner was Marta Cerruti and he asked her if she could put Havens in touch with her father, the mathematician Umberto Cerruti from the University of Turin. Because his daughter had asked so kindly, Cerruti decided to help the writer of the letter. He sent him a mathematical problem to solve. And so it happened: Havens replied with a long difficult formula, on a sheet of paper measuring 1 and 20 cm long. Cerruti checked it on his computer, and it turned out to be correct.

Then Cerruti sent Havens a question about continued fractions, something Cerruti was researching for himself. After several discussions by airmail a top scientific article was born. Cellmates of Havens found inspiration in his story, and now there is a math club in prison.

On 14 March (also written as 3.14) each year they celebrate Pi Day behind bars. Havens sees his research as a way to pay off his debt to society and, as Cerruti explained, 'he always dedicates his mathematical findings to his victim.'

Havens' Continued Fractions

Today mathematicians represent a continued fraction as follows:

$$z + \cfrac{y}{x + \cfrac{w}{v + \cfrac{u}{t + \cdots}}}$$

The principle of a fraction with an infinite number of fractions continuing in the denominator dates back to the times of Euclid, although fractions were then simply seen as a division of natural numbers. The greatest mathematicians, such as Leonhard Euler, Joseph-Louis Lagrange, Carl Friedrich Gauss, Évariste Galois, André Weil and Srinivasa Ramanujan would work on it. Galois and Weil also did their mathematics partly in captivity.

For example, a continued fraction for the number pi = 3.141592..., the circumference of a circle divided by its diameter, is

$$\pi = 3 + \cfrac{1}{7 + \cfrac{1}{15 + \cfrac{1}{1 + \cfrac{1}{292 + \cfrac{1}{1 + \cdots}}}}}$$

This can be verified by considering the following successive approximations:

$$3 + \frac{1}{7} = \frac{22}{7} = 3.142857\ldots,$$

$$3 + \cfrac{1}{7 + \cfrac{1}{15}} = 3.141509\ldots,$$

$$3 + \cfrac{1}{7 + \cfrac{1}{15 + \frac{1}{1}}} = 3.141592\ldots, \ldots$$

There is no regularity, repetition or pattern in the numbers 3, 7, 15, 1, 292, 1 etc. However, with another arrangement this is indeed the case:

$$\pi = \cfrac{4}{1+\cfrac{1}{3+\cfrac{2^2}{5+\cfrac{3^2}{7+\cfrac{4^2}{9+\cdots}}}}}$$

Finding such beautiful expressions is not easy. It is a difficult research field, which today mainly finds application in networks.

'Held Back' from Congresses

Whether Havens will accomplish as many great things as the Frenchman André Bloch (1893–1948) remains a question. Bloch wrote a complete oeuvre in captivity. He spent his life in the psychiatric institution Saint-Maurice, because at the age of 23 he had stabbed his brother, uncle and aunt during a family meal, 'to wipe out a mentally ill family forever' (Fig. 6.2).

A psychiatrist stated that Bloch was suffering from 'morbid rationalism', which would be as dangerous as the more famous 'irresistible impulse', a totally opposite 'affliction'. Others claim that the cause was an injury to the prefrontal cortex, which Bloch sustained in World War I. Still others saw the murder as evidence that the family was mentally ill. After the murders, Bloch ran to the street to surrender without resistance. For the rest of his life, he practiced math, as a model internee.

> 'For forty years this man sat at a table every day in a small hallway leading to his room, and he never moved from his place, except to eat, until evening. He spent his time writing algebraic equations and mathematical signs on scraps of paper or reading and annotating books on mathematics whose intellectual level was that of the great specialists in the field. [...] At 6:30, he closed his notes and books, ate, immediately returned to his room, fell on his bed, and slept until the next morning. Where other patients constantly asked for freedom, his equations and his correspondences were sufficient for him.'

Bloch had a subscription to the *Bulletin des Sciences Mathématiques* and regularly wrote to top mathematicians such as Jacques Hadamard, Gösta Mittag–Leffler, George Pólya and Henri Cartan. He often dated his letters 1 April, even though they were written on a different day. And it happened

UN TRIPLE ASSASSINAT

Drame de la folie

Un terrible drame s'est déroulé hier, entre midi et une heure, 104, boulevard de Courcelles.

A cette adresse habitait M. Charles Didisheim, âgé de quarante-six ans, et Mme Didisheim, née Linda Olivetti, trente-trois ans, fabricants de montres.

Les époux avaient invité à déjeuner leur frère, M. Georges Didisheim, lieutenant au 83e régiment d'artillerie, détaché à l'Ecole polytechnique, et leur neveu André Bloch, lieutenant au 117e d'artillerie lourde, demeurant 26, avenue de Wagram.

Le lieutenant Bloch était en convalescence de deux mois par suite d'une commotion cérébrale reçue au cours d'une attaque sur l'Aisne. Il n'était pas complètement remis et accusait, depuis sa blessure, des dérangements cérébraux fréquents, mais rien ne faisait présager le drame dont il allait être l'inconscient acteur.

Comment le tragique événement s'est-il déroulé ? On l'ignore.

Ce que l'on a appris c'est que le lieutenant a tué à coups de revolver son oncle, sa tante et son cousin dans la salle à manger, où ils déjeunaient tranquillement.

Le fou s'est ensuite acharné sur les cadavres à coups de couteau.

L'oncle porte les traces d'une cinquantaine de blessures.

Quand il fut las de frapper, le lieutenant Bloch se rendit au commissariat de police de la Plaine-Monceau, fit mander M. Raoul Legrand, commissaire de police, et lui dit :

— Poursuivi par la société d'une haine injuste, je viens de me faire justice, je les ai tous tués.

Et il expliqua son triple assassinat.

M. Legrand se rendit boulevard de Courcelles et ne put que constater que le malheureux avait dit vrai.

Le lieutenant Bloch a été écroué à la prison du Cherche-Midi.

A TRIPLE MURDER

Drama of madness

A terrible tragedy took place yesterday between noon and 1 PM, 104 Boulevard de Courcelles.

At this address lived Mr. Charles Didisheim, forty-six years old, and Mrs. Didisheim, born Linda Olivetti, thirty-three, watch manufacturers.

The couple had invited their brother, Mr. Georges Didisheim, lieutenant in the 83rd Artillery Regiment, detached from the École Polytechnique, for lunch, and their cousin André Bloch, Lieutenant in the 117th Heavy Artillery, residing at 26, avenue de Wagram.

For two months, Lieutenant Bloch was recovering from a concussion sustained during an attack on the Aisne. He had not fully recovered and had frequent brain disorders since his injury, but nothing suggested the drama in which he would become the unconscious actor.

How did the tragic event unfold? We do not know.

What we did hear is that the lieutenant shot his uncle, aunt and nephew in the dining room, where they were quietly having lunch.

Then the madman stabbed the corpses with a knife.

The uncle bears traces of about fifty wounds.

When he was tired of stabbing, Lieutenant Bloch went to the Plaine-Monceau police station, sent for Mr. Raoul Legrand, Police Commissioner, and said:

- Chased by society with unwarranted hatred, I just did myself justice, I killed them all.

And thus he explained his triple murder.

Mr. Legrand went to Boulevard de Courcelles and could only note that the unfortunate person had told the truth.

Lieutenant Bloch was imprisoned in the Cherche-Midi prison.

Fig. 6.2 Paper of 18 November 1917 about Bloch's murders, in the French newspaper *Le Figaro*

that he had to apologize to his colleagues for his absence from a mathematics conference because 'he was held back'.

Bloch is mainly known for complex analysis, which is the theory of functions with complex numbers (numbers that include the imaginary square roots of −1). Today, 'Bloch's Theorem' (1925) is still in the handbooks. There is even a number named after him, Bloch's constant. For all this groundbreaking mathematical work, the Prix Becquerel of the French Académie des Sciences was awarded to the triple murderer in 1948, just before he died, as a scientific Last Sacrament (Fig. 6.3).

Fig. 6.3 This figure represents the complex function $f(z) = (z^2 - 1)(z + 2 - i)^2/(z^2 + 2 - 2i)$; where $z = x + iy$ where i is an imaginary square root of -1. The colour and brightness provide information about $f(z)$

7

Murdering Emperors

A team around Brazilian mathematician F. A. Rodrigues improved a 'survival analysis' of the 69 Roman emperors by American J. H. Saleh. It used a so-called Pareto distribution, known from the 80/20 rule from computer, Netflix, or food habits. Can this survival analysis also be applied to current local elections too?

Fig. 7.1 The first Roman emperor, Augustus, died peacefully after the longest-running reign in imperial history (left, Augustus as Jupiter, The Hermitage, St Petersburg, Russia). Emperor Galba was assassinated after being betrayed by his own Praetorian guards, though they were set up to protect the emperor (right, statue of Galba by Bartolomeo Cavaceppi, Sala Rotonda Museo Pio-Clementino, Rome, Italy)

© The Author(s), under exclusive license to Springer Nature
Switzerland AG 2023
D. Huylebrouck, *Dark and Bright Mathematics*, Copernicus Books,
https://doi.org/10.1007/978-3-031-36255-2_7

Mathematical Intrigues

In 44 BC, Julius Caesar (100–44 BC) was appointed 'dictator perpetuus', but he was murdered after a month by, among others, his own (adopted) son Brutus. Though he was not an 'emperor' himself, Caesar began a long line of Roman emperors. The first actual Roman emperor was his grandnephew Octavian, better known as Augustus. He took power after a series of internal struggles (from 27 BC to 14 AD, although some date the beginning as 31 BC). He would die of natural causes, like his successor, Tiberius, although the latter may have been suffocated on his sickbed with a pillow. After them, most Roman emperors would die in a way that is reminiscent of the series *Game of Thrones*.

The third emperor, Caligula, was killed after barely four years in power by his own Praetorian Guard, who were supposed to protect him. Next, Claudius lasted thirteen years before being poisoned. The following emperor, Nero, lasted nine months more before committing suicide, 'aided' by his secretary. Galba then reigned for seven months, until he fell into the hands of mutinous soldiers in the Roman Forum after his Praetorian Guard had betrayed him. The instigator of this murder became the new emperor, Otho, for three months, until he committed suicide. Then it was Vitellius' turn, who after eight months was dragged to the forum with a noose around his neck and pelted with dung and rotten fruit, ending up in the Tiber after being tortured. Vespasian broke through this cruel series and lasted ten years, to die finally from illness. His son, Emperor Titus, died of a fever after just two years, though some say he was poisoned (see Fig. 7.1).

And so it goes on with the deaths of the more than 70 Roman emperors: three quarters died by assassination, poisoning, in battle or suicide (their number varies slightly depending on whether regencies or dubious deaths are included). In 395 AD the Roman Empire split into two parts, but the numbers barely improve when the Eastern Roman Empire is included, as then one in three of its 175 emperors did not die peacefully. Coincidentally or not, the last Byzantine emperor also died in battle: Constantine XI Palaiologos threw off his imperial cloak to avoid capture and to better fight the Ottomans but was thus killed, in 1453.

The history of the Roman emperors is a web of intrigues of the Praetorian Guard, of the leaders of the legions, of the Senate and of their own families too. And so, an emperor reigned on average for only 5.6 years. And yet, in 2021 P. L. Ramos, F. Louzada and F. A. Rodrigues, of the Institute of Mathematics and Computer Science, and L. F. Costa of the Institute of Physics at the University of São Paulo in São Carlos, Brazil, managed to characterize these turbulent historical events by mathematical laws.

Survival Statistics

Lead researcher Rodrigues and his colleagues built on a 2019 publication by American Joseph H. Saleh, of the Georgia Institute of Technology in Atlanta, USA. Saleh was the first to apply so-called survival analysis to the lives of the Roman emperors. This branch of statistics studies the expected time until an event occurs, such as the death or illness of persons or beings (in medical sciences), failure of mechanical systems (in engineering), unemployment after job loss, and so forth.

The survival function $S(t)$ is the probability that the being of study lives longer than a given moment t, or that the object of study functions longer. It is often approximated using the so-called 'Kaplan–Meier estimator', named after American mathematician Edward L. Kaplan (1920–2006) and biostatistician Paul Meier (1924–2011). Their formula gives a simple estimate for survival function at a time t.

Suppose that d_i is the number of deaths in a time span t_i when n_i beings are still alive, or the number of defective machines in a time span t_i when n_i devices are still working. Then $1 - d_i/n_i$ is the number of survivors in that time span t_i. So, if we multiply all those numbers for all time spans t_1, t_2 ... to t_i, which together add up to the time span t, then this product will be an estimate of the survival function at a time t. It is the Kaplan–Meier estimator:

$$S(t) = (1-d_1/n_1).(1-d_2/n_2). \ldots .(1-d_i/n_i)$$

Saleh thus calculated that of the 69 emperors he considered (he ignored some doubtful cases), 11 lasted about half a year before being assassinated. So that is 1–11/69 or about 84% who lasted longer, although two also died of natural causes within six months. Thus, 56 emperors remained. Of these, five died from violence within the year, and thus the value of the survival function now becomes: $(1-11/69).(1-5/56) = 76\%$. The calculations continue, to result in a typical Kaplan–Meier step graph, with percentages of 84%, 76%, and so on (see illustration). For example, the figure shows that after four years (on the mark of Caligula) an emperor had a 60% chance of not having met a violent death, or conversely, a 40% chance of having died violently. If he lasted as long as Augustus, he would have had a 24.8% chance of not dying violently, or a 75.2% chance of being murdered (Fig. 7.2).

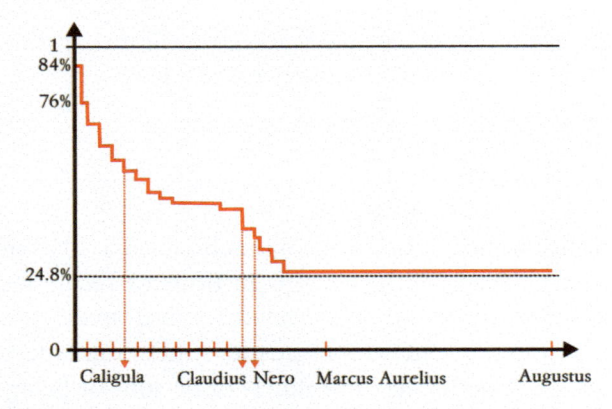

Fig. 7.2 Kaplan–Meier step graph for the death of Roman emperors

Saleh now proposed a Weibull curve to approximate the shape of those steps, because the Swedish mathematician Waloddi Weibull (1887–1979) had successfully used them to model lifetime distributions in the classic examples of medical experiments or mechanical devices. It has the shape of a (decreasing) exponential curve, today mainly known for (decreasing) coronavirus figures, but multiplied by the variable itself. Two more parameters are also involved, although they are not called a 'reproduction number', as with the coronavirus curves, but shape and scale parameters. Saleh therefore looked for the most suitable Weibull curve, although it turned out that he only needed the following combination:

$$S(t) = 0.876 \cdot e^{-(t/12.835)^{0.618}} + 0.124 \cdot e^{-(t/14.833)^{13.387}}$$

From that equation, and more particularly from the number 0.618, Saleh concluded that Roman emperors were subject to 'infant mortality'. This technical term from the survival analysis indicates someone dies or becomes ill shortly after the treatment in the study, or that a machine breaks down shortly after it is put into operation. In the case of the Roman emperors, it seemed that they died very soon after their accession to the throne. Was this due to lack of experience or because throne changes were simply turbulent?

If the emperors survived the initial period unscathed, their chances of survival increased, but then, according to Saleh, the number 13.387 implied the emperors also suffered 'wear-out mortality'. Again, that is a technical term that indicates death due to old age or being out of function due to wear and tear. In the case of the emperors, it meant that that after 12–13 years they were again at greater risk of being murdered. Was this due to the apostasy of their allies or the zeal of new rivals? (Fig. 7.3)

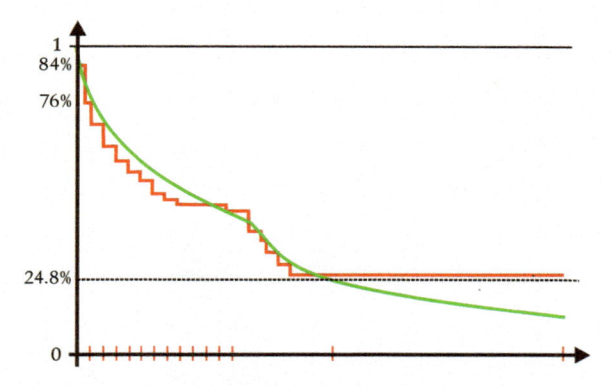

Fig. 7.3 Saleh's approximate Weibull curve

Power Curves

Saleh's conclusion could also have been derived from the steep descending steps in the beginning and after Claudius or Nero, so that a non-mathematician might wonder what all that calculational muscle with those Weibull curves was good for. This, however, was well illustrated in another publication, in the *Royal Society Open Science*, a peer-reviewed scientific journal of the British Royal Society, in which the group around Rodrigues abandoned the Weibull description, common in statistical survival studies, to opt for so-called 'power curves'. Here, the word does not refer at all to any political connotation, but simply to that of 'raising to a certain power'. After all, these functions are given by a rule of the form x^a, in which x is raised to the power a. If $a = 2$ then this corresponds to raising to the second power, that is, to squaring. If $a = 3$, to a 'third power', etc. If $a = 1/2$ it corresponds to a square root.

Rodrigues also needed a combination of two curves to approximate the Kaplan–Meier steps:

$$S(t) = 0.248 + 0.752 \cdot \left(\frac{t}{0.5}\right)^{-0.382} 1_{[0.5, 13]} + 0.522 \cdot \left(\frac{t}{13}\right)^{-7.5} 1_{[13, \infty]}$$

Rodrigues and his colleagues were able to show that their power law fits the data statistically better than the Weibull curve. That too can be seen with the naked eye: not only does it better show the abrupt change at 12–13 years mark, their curve also does not continue to go downwards on the right (Fig. 7.4).

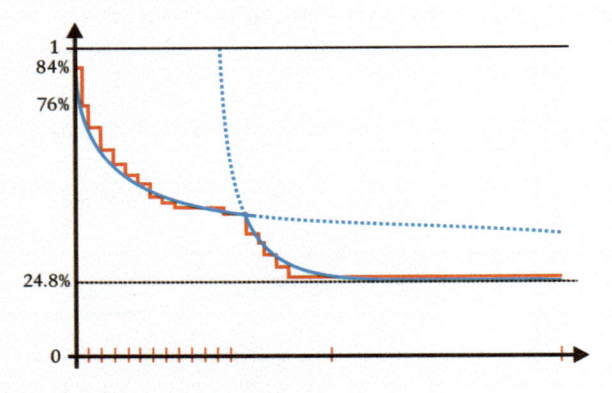

Fig. 7.4 Rodrigues' approximating power curves

These power curves are also called Pareto distributions, after the Italian economist Vilfredo Pareto (1848–1923). In 1906, he studied the distribution of wealth and found that 80% of Italy's property belonged to 20% of the population. His 80/20 rule aptly translated how the majority had few resources and a minority owned most of the wealth.

This rule subsequently appeared in a large number of applications: 80% of computer users use only 20% of the computer power; 80% of all people only watch 20% of Netflix's content; 80% of a text consists of the 20% most frequently used words; 80% of the lunar craters are small, 20% large; 80% of social media accounts have few followers, 20% have many. Even diet gurus use it: 'Eat healthy 80% of the time, then you can sin 20% of the time'. Managers state that '80% of the turnover is earned thanks to 20% of the customers'; one should therefore pamper that 20%.

In coronavirus times, this 80/20 principle sounded like an often-heard slogan: '20% of the people are responsible for 80% of the trouble!' But there is truth in it: if a phenomenon can be described by a certain power curve, then 20% of the causes determine 80% of the effects. And that is not a simplification of reality (though in some cases it is, of course), because a mathematical law, namely the Pareto probability distribution, is at the basis of it. In the case of Roman emperors, the rare occurrence was the peaceful death, and the most common occurrence was being murdered. The numbers are slightly different because the Pareto power curve has different parameters, but the principle is the same.

Meaningfulness

Some historians will, of course, doubt the usefulness of applying mathematical methods to historical phenomena, but Saleh was prepared for this criticism:

> 'Beyond these specific details, what does it mean to find a coherent structure within a stochastic process of historical nature as the one here examined? Roughly speaking, the result implies the existence of systemic factors and some level of determinism [...] superimposed on the underlying randomness of the phenomenon here examined. In other words, the process is not completely aleatory; it has some deterministic factors overlaid on its randomness. Conan Doyle, in Sherlock Holmes: The Sign of Four, expressed this general idea rather accurately when he wrote:
> 'While the individual man is an insoluble puzzle, in the aggregate he becomes a mathematical certainty. You can, for example, never foretell what any one man will do, but you can say with precision what an average number will be up to."

Rodrigues' discovery that power functions are better suited to describe Roman imperial deaths motivates the use of mathematical methods in the study of history. After all, why would the application of the 80/20 rule to history make less sense then for describing computer usage, Netflix, preferred words, or the size of lunar craters?

Of course, historians will comment that statisticians slightly manipulate some historical facts. What to do, for example, with the case of Emperor Hostilianus, who became emperor at the age of twenty but died of a disease five months later and could therefore not be murdered? Such studies introduce simplifying premises, but that is often the case in mathematical applications, including in accepted applications in physics or in astronomy.

Another question is whether this research could have further applications in other eras or situations. One might, of course, immediately think of Alexander the Great, who also reigned for thirteen years, from 336 to 323 BC, until he died a suspicious, at least early, death. That length of time is curiously close to the turning point in the Roman Empire. Note researchers already established that the duration of the regimes of 106 dictators between 1875 and 2004 was on average 12.3 years—just think of Adolf Hitler, who came to power in 1932 and committed suicide in 1945.

Applicable Today?

One can therefore wonder whether this survival analysis could also be applied to local situations of today. Rulers are no longer killed or forced to commit suicide, but they can lose power after elections or an internal party struggle. Here is a list of prime ministers of the government of the Flemish state, Flanders, which is a part of Belgium, and of prime ministers of the federal government of Belgium. When a prime minister led a government interrupted by someone else, he got a roman number behind his name. And so, when one of them, Yves Leterme was first head of the Flemish government, and later twice of the Belgian government, he was first called Leterme I, and later Leterme II and III. The data begins within the lifetime and memory of the author, so that he could make a judgement about why those prime ministers left power without having to do historical investigations (Fig. 7.5).

Perhaps the reader can make up a similar list for the state governors or of town mayors she/he lives in and for her/his (federal) government. This table is merely an example of how the survival analysis for the Roman emperors could be applied to a current local situation (Fig. 7.6).

These data lead to a new table and to the typical Kaplan–Meier step graph, with the calculations given here in their entirety, so that each reader can adapt them according to his personal situation (Fig. 7.7).

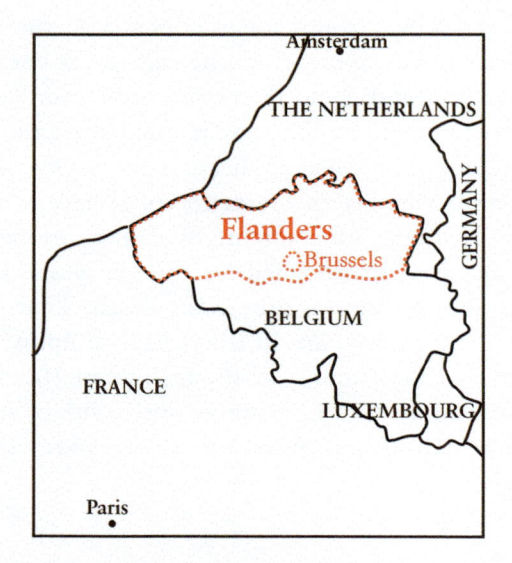

Fig. 7.5 Flanders is a part of Belgium

Prime Minister Government Flemish State		Begin power	End power	Years power	Forced to leave power?	Motivation
First name	Surname					
Gaston	Geens	1981	1992	11	No	Retired
Luc	Van den Brande	1992	1999	7	Yes	Election defeat
Patrick	Dewael	1999	2003	4	No	Retired
Bart	Somers	2003	2004	1	Yes	Election defeat
Yves	Leterme I	2004	2007	3	No	Became Prime Minister of Belgium
Kris	Peeters	2007	2014	7	Yes	Election defeat
Geert	Bourgeois	2014	2019	5	Yes	Internal party decision
Prime Minister Government of Belgium						
Leo	Tindemans	1974	1978	4	Yes	Government fell
Paul	Vanden Boeynants	1978	1979	1	No	Retired
Wilfried	Martens I	1979	1981	2	Yes	Government fell
Mark	Eyskens	1981	1981.5	0.5	Yes	Government fell
Wilfried	Martens II	1981	1992	11	Yes	Election defeat
Jean-Luc	Dehaene	1992	1999	7	Yes	Election defeat
Guy	Verhofstadt	1999	2008	9	Yes	Election defeat
Yves	Leterme II	2008	2008.5	0.5	Yes	Government fell
Herman	Van Rompuy	2008	2009	1	No	Went to the EU
Yves	Leterme III	2009	2011	2	Yes	Election defeat
Elio	Di Rupo	2011	2014	3	Yes	Election defeat
Charles	Michel	2014	2019	5	No	Went to the EU
Sophie	Wilmès	2019	2020	1	Yes	Internal party decision

Fig. 7.6 List of recent rulers in Flanders and Belgium

Years power	Number remaining	Under pressure	Own choice	Kaplan–Meier estimator
0.5	20	2	0	1−2/20 = 0.90
1	18	2	2	0.90.(1−2/18) = 0.80
2	14	2	0	0.80.(1−2/14) = 0.6857
3	12	1	1	0.6857.(1−1/12) = 0.6286
4	10	1	1	0.6286.(1−1/10) = 0.5657
5	8	1	1	0.5657.(1−1/8) = 0.4950
6	6	0	0	0.4950.(1−0/6) = 0.4950
7	6	3	0	0.4950.(1−3/6) = 0.2475
8	3	0	0	0.2475.(1−0/3) = 0.2475
9	3	1	0	0.2475.(1−1/3) = 0.1650
10	2	0	0	0.1650.(1−0/2) = 0.1650
11	2	1	1	0.1650.(1−1/2) = 0.0825

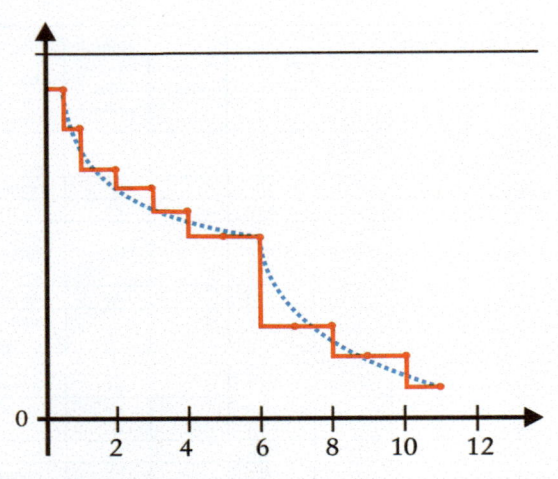

Fig. 7.7 Reworked table and Kaplan–Meier step graph for Flanders and Belgium

As with the Roman emperors, it turns out there is an infant mortality and a remarkable turning point, not around 13 years, but six. And only 30% of the leaders left power by their own will. There seems no point in drawing further conclusions or fitting a mathematical curve to those data, due to their insufficient number. A much greater amount of data could follow from the study of the hundreds of changes in power in cities and villages. However, this requires the input of politicians or political scientists. It is suspected, however, that they must first be convinced of the usefulness of the herein proposed survival analysis, since their own survival does not yet depend on it.

8

When the Dead Talk in Code

After deciphering messages from Zodiac Killer and Langrenus, Belgian code specialist Jarl Van Eycke focused on cryptic message figuring on altars in a church in Moustier, near the border of Belgium and France. Does this cryptogram contain clues to a Templar treasure?

Fig. 8.1 The altars of the two aisles of the main church of Moustier have cryptic texts. The signs appeal so much to the imagination the US Secret Service devoted an article to it

© The Author(s), under exclusive license to Springer Nature Switzerland AG 2023
D. Huylebrouck, *Dark and Bright Mathematics*, Copernicus Books,
https://doi.org/10.1007/978-3-031-36255-2_8

After the Zodiac Killer and Langrenus, an Altar Code?

Moustier, a borough of Frasnes-lez-Anvaing in the province of Hainaut, Belgium, is recognizable from afar by its two towers standing next to each other: one of a church, the other of a very large chapel just next to it. They raise an intriguing question: where did the humble rural village get the money to build two churches next to each other? (Figs. 8.1 and 8.2).

What really matters here, however, are the inscriptions that can be found in the church. Inside the building are two marble altars. One is dedicated to Saint Martin, the other to Mary. The pedestals of the altars are decorated with a kind of Tables of Moses. They contain Latin and Greek characters as well as some cryptic characters.

For decades, no one has been able to decipher them. And so, the code became part of top ten lists of remarkable secrets. The NSA, the American National Security Agency, devoted a paper to it in the September 1974 issue

Fig. 8.2 Two church towers, close to each other, adorn the view on the village of Moustier

of its magazine *Cryptolog*. It was entitled 'Secrets of the Altars'. In the nearly fifty years that passed since then, no one has proposed a decipherment.

Evelyn Bastien, a veterinarian from the nearby village Cambron-Casteau and volunteer Latin teacher, has been fascinated by the Moustier codes for years. When she read another paper of mine, about the deciphering of the code of Belgian astronomer Langrenus, she contacted me, hoping that the Moustier mystery could be cracked in a similar way. At our request, Jarl Van Eycke had pondered over the more than 400-year-old Langrenus code. In less than two weeks, this Belgian genius cracked the code that had left so many puzzled. It was not a first for Van Eycke: at the end of 2020 he had helped to decode the mysterious message of the Zodiac Killer, on which the FBI had been struggling for more than half a century. 'What is Van Eycke's opinion on the Moustier code?' Bastien wondered.

Cryptography or Abbreviations?

In 1838 the church of Moustier was restored. The parish registers mention that 'a stonemason received board and lodging for eighteen days'. That stonemason would have been Pierre Brébart, from the nearby town of Tournai. This is written in a fifty-year-old booklet, published in-house by a resident of the village, Jean Connart. Connart paid attention to the cryptograms and went looking for a solution. He found no information in the church archives. The closest he came to a possible decipherment was by a chance meeting with a certain Paul de Saint-Hilaire. In the 1970s, this award-winning writer focused on esoteric subjects, such as the Templars.

The writer pointed Connart to the coding method of the German Benedictine abbot Johannes Trithemius (1462–1516), author of the *Polygraphia*, the oldest printed work on cryptography. It appeared in 1518 and received a great deal of attention in numerous brotherhoods. In Moustier, a 'Confrérie à la Vierge Marie' was founded in 1488, an association of laymen committed to the Church. Trithemius' book contains endless pages in which one letter of the alphabet is associated with one word. Whoever owned the book could read texts coded according to his method. In his book, Connart included the complete Latin text of the deciphered altar codes but made no translation of it. Because, Connart thought, 'one cannot attach meaning to it'.

That is not so. A legible text does appear, with a religious meaning. Let's show how to decipher the beginning of the codes according to Trithemius' *Polygraphia*. The codes on the altar start with the letters JNLK. The first column of the many pages of codes states that I (the Latin J) corresponds

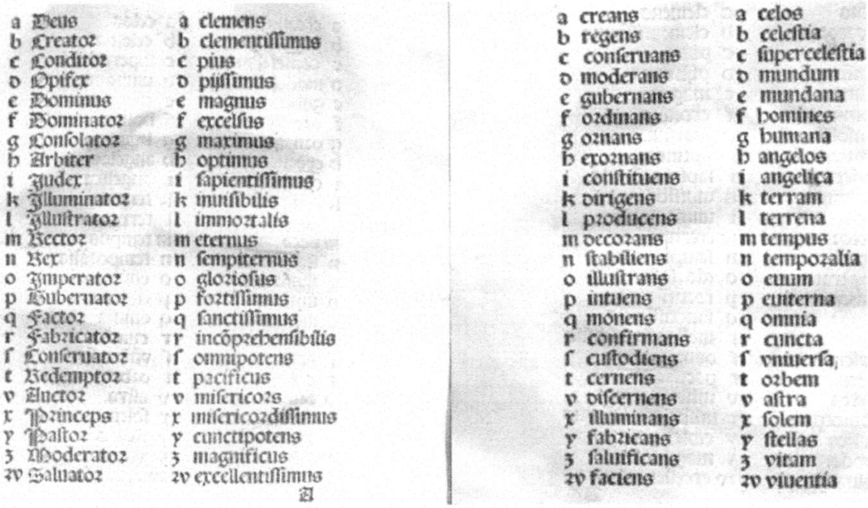

a Deus	a clemens	a creans	a celos
b Creator	b clementissimus	b regens	b celestia
c Conditor	c pius	c conseruans	c supercelestia
d Opifex	d piissimus	d moderans	d mundum
e Dominus	e magnus	e gubernans	e mundana
f Dominator	f excelsus	f ordinans	f homines
g Consolator	g maximus	g ornans	g humana
h Arbiter	h optimus	h exornans	h angelos
i Iudex	i sapientissimus	i constituens	i angelica
k Illuminator	k inuisibilis	k dirigens	k terram
l Illustrator	l immortalis	l producens	l terrena
m Rector	m eternus	m decorans	m tempus
n Rex	n sempiternus	n stabiliens	n temporalia
o Imperator	o gloriosus	o illustrans	o cuium
p Gubernator	p fortissimus	p intuens	p cuncterna
q Factor	q sanctissimus	q monens	q omnia
r Fabricator	r incomprehensibilis	r confirmans	r cuncta
s Conseruator	s omnipotens	s custodiens	s vniuersa
t Redemptor	t pacificus	t cernens	t orbem
v Auctor	v misericors	v discernens	v astra
x Princeps	x misericordissimus	x illuminans	x solem
y Pastor	y cunctipotens	y fabricans	y stellas
z Moderator	z magnificus	z saluificans	z vitam
w Saluator	w excellentissimus	w faciens	w viuentia

Fig. 8.3 The first pages of codes from Trithemius' *Polygraphia*

to 'Luder', while the second shows that N is 'sempiternus', and so on. And so: 'Luder sempiternus produces terram' or: 'The eternal player producing the earth...'. The beginning on the altar of Mary is LGammaEG, or: 'Illustrator ? gubernans humana...', that is: 'The enlightened one ? who rules over man...'. We replaced the Greek gamma, which is not in the second column, with a question mark, but this column only contains adjectives so that this word has little influence on the actual content. In this way, a text about the goodness of God appears on both tables (Fig. 8.3).

Il?lustris Karolus (B ?) PRaefectus Vir Magnificus Gratis Haec VoVit Ecclesiae Qui Legitis Sit Benedictus Nostro ? PaPa MaGister ? Karolus, Honoris Usum Remicet ? Libertissime Reddidit Nobis ? Solidos XV	The Exalted Charles (B gamma), administrator and great man, has selflessly given these things to the Church, he who is blessed by the Pope Charles Lemaître has bestowed the honour, very generous, to pay us 15 cash

That is, of course, quite possible for a text in a church. Moreover, the signs on the tables of the altars correspond nicely with two pages of the *Polygraphia*. This may be an indication of the use of this kind of encryption that was in vogue at the time. But, on the other hand, any sequence of letters according to the system of Trithemius leads to a religious text. And the content doesn't indicate a need for encryption: no real secrets are revealed.

Connart also suggested the altars could consist of classical abbreviations. Think, for example, of the well-known 'SPQR', the abbreviation of 'Senatus Populusque Romanus' (The Senate and People of Rome). It is on many Roman monuments. The Church had its abbreviations too. Perhaps the altars are lists of gifts? According to a lexicon of abbreviations, the *Lexicon Abbreviatorum* (1901) by Adriano Cappelli, the left part of the altar of Saint Martin could read:

A certain Magister Karolus (1693–1778), whose name could be translated as Charles Lemaître, is indeed buried just in front of the altar of Saint Martin, with his family. Two of his sons were priests, a daughter was a nun and a grandson a canon. Another grandson, also carrying the first name Charles, lived during the construction of the altars. It is likely that these donations were coded in a church, though it raises new doubts. After all, the decoding is very speculative, with many question marks. And donors of other gifts, such as for the stained-glass windows, are mentioned in the church with names and titles (Fig. 8.4).

Fig. 8.4 This stained-glass window was donated by the Miss Augusta du Sart, and it clearly says so

Echo of the Merovingians

In the footsteps of his uncle, Philippe Connart also devoted himself to the history of Moustier. He looked more specifically at the forms of the letters in the cryptograms. They vary, although both tables clearly seem to have been sculpted by the same person. It is striking that the text contains the symbols J, U and W, as these letters do not appear in the Latin alphabet. Neither do Y and Z, because they were only added later for the transcription of Greek words. Latin was nevertheless the official language in the Church until the Second Vatican Council (1962–1965). The inverted L perfectly resembles a Greek capital gamma. An inverted V also occurs a few times; is that the Greek capital letter lambda? The letters were colored in black, as is often the case with texts carved on stone, to increase legibility. But here and there, distinctly chiseled dashes were skipped without even a hint of a remnant of black.

According to Philippe Connart, the stonemason based the curious shape of his letters on signs from the tenth century, used by a monk of the abbey of Saint-Amand-les-Eaux. That village is now located in France, on the border with Belgium. It is barely 40 km from Moustier. These 10th-century Merovingian letters, in turn, are said to have been derived from 4th-century works by a bishop of Constantinople, Saint Gregory the Theologian, founder of the Trinity doctrine. If those time lapses seem far-fetched, Connart's main argument certainly is, as it is all about the resemblance of just one letter, a kind of angular C. That type of letter can be found across all those centuries. Overall, Connart did provide compelling evidence of Moustier's ties to the distant past, but they didn't help him to decipher the code.

Magritte's Surrealism

We leave the trail of Jean and Philippe Connart and return to the Templars. Twenty years ago, Rudy Cambier, who studied languages at the University of Liège, wrote a book about it: *Nostradamus and the legacy of the Templars*. In the book, he proposes a new hypothesis about the Moustier code. According to Cambier, it was written in a consonant script, in which the reader must fill in appropriate vowels. The ten lines would refer not only to the ten commandments from the Bible but also, in one fell swoop, to a hidden treasure of the Templars. To this end, the linguist reinterpreted the book *Prophecies* by the 16th-century astrologer, apothecary, and physician Michel Nostradamus.

According to him, it was based on an older book, *Centuries*, by the 14th-century Yves de Lessines, a monk from the abbey in nearby Cambron-Casteau (just next to where Pairi Daiza zoo is today) (Fig. 8.5).

Nostradamus' book contains words in Picardian, Moustier's dialect, Cambier claims. Moreover, the astrologer is said to have visited the region and the word 'moustier' is indeed mentioned in Nostradamus' writings, in the chapter Century I, verse 95:

'Devant moustier trouvé enfant besson,
 D'heroic sang de moine & vetustique…'

The old French word 'besson' means 'twins', which can refer to the two churches of Moustier, or to the two altars. Or, according to Cambier, to two holy water fonts. One of them is in the church and the second one would be at his home, in nearby Wodecq, where there would also be a Templar cross on a beam. All that would be a reference to a hidden treasure of the Templars, on the property of Cambier! (Fig. 8.6)

Others strongly doubt whether Nostradamus ever visited the Moustier region. Moreover, the reference to the village of Moustier in his verses is not a certainty, since 'moustier' is an old French word for 'monasterium'. In an unbiased translation by Marten Hofstede, there is no mention of the Walloon village:

'An old monk finds a twin child of
 Heroic blood in front of a monastery.'

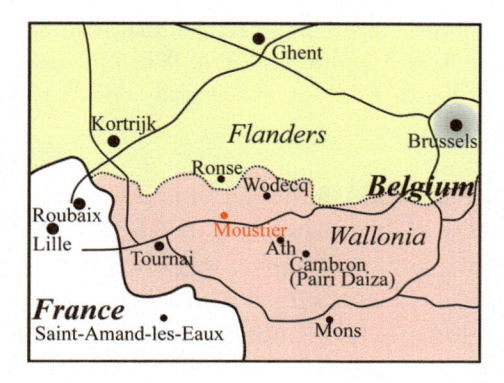

Fig. 8.5 Southeast of Moustier lays Cambron, with the Cistercian abbey where the monk Yves de Lessines lived

Fig. 8.6 The Templar's Cross in the church

Some soil probes on Cambier's site did indicate underground anomalies, but in 2001 the Walloon government refused to allow excavations. These were done on an adjacent terrain, until a land subdivision permit was granted there. With the argument that 'routes pass through this field that are described in the *Centuries*', it could be halted. So don't say any longer Nostradamus cannot be used for one's own benefit.

The then mayor of neighboring village Frasnes-lez-Anvaing and later foreman and minister of the Belgian political party MR Jean-Luc Crucke argued:

> 'I would very much like the site to be classified as a Walloon heritage [...] With this treasure we find ourselves in the surrealism that is already strongly present in our region where René Magritte was born. When the treasure is discovered, that is the icing on the cake. If not, and it really doesn't matter, it appeals to the imagination.'

A treasure was never found, although unknown persons did damage the stones under the altars in their search (Fig. 8.7).

Philippe Connart, mentioned above, was one of the few who could see Cambier's decipherment. He described it as follows: 'It takes a lot of imagination to read anything into it.' Cambier didn't let it bother him: 'I just want to expose Nostradamus and bring the excellent author Yves de Lessines to wider attention'.

Fig. 8.7 Damaged stones under the altars

Contemporary Opinion

In the modern computer age, genius decipherers might be able to see through the code in no time, Bastien thought. In particular, she put her hopes in Jarl Van Eycke. And so, we did present him with the cryptograms. Van Eycke quickly pointed out the lack of repetitions per line. For the altar of Saint Martin, there are only seven lines in which a letter is repeated. And when the code is cast in seven-character lines, he went on to say, the number of repetitions should normally drop, but it rises, to eleven. He checked that even with lines of six characters each, there are still nine repetitions in total. According to Van Eycke:

> 'This means that when the code was created, there was a conscious intention not to repeat too many characters per line. From a cryptographic point of view, this seems strange to me.'

British computer specialist and writer Nick Pelling had also pondered on the code on an internet forum. According to him, the letters of the cryptograms do not look as if they were chiseled by a good craftsman. The marble stones could therefore simply have been some practice material on which a stonemason had carved some letters. That meaningless text resembling tables of the Ten Commandments could have been incorporated as a stopgap to adorn the altar. And after all, that part is usually covered by a cloth.

Bastien was disappointed. Why would a parish that could build two churches barely 10 m apart not have had the money for a good stonemason—and, more than that, for the most important part of each church, the altar? Moreover, the aforementioned Pierre Brébart was indeed a professional sculptor, who could not have carved just anything in the eighteen days during which he received board and lodging. And so, there remains a mystery in Moustier.

Part II

Down-to-Earth Math

9

Columbus's Reference Line on Earth

Today Oradea is a quiet town in Romania, but in the time of Christopher Columbus it was the geographic reference for the whole world—a bit like Greenwich is today.

Fig. 9.1 On the Caverio map (1505), named after its author, cartographer Nicolay de Caveri from Genoa, the Balkan region is indicated by a tent (triangular-shaped, right above the boot of Italy), symbolizing the Ottoman occupation

Georg von Peuerbach

The Romanian town of Oradea, or Nagyvárad in Hungarian, or Großwardein in German, is located in the northwest of the country, about ten kilometers from the Hungarian border. As its different names suggest, the town has a rich history: the earliest known settlement dates from the end of the Neolithic period. In 106 BC the Romans conquered most of the Dacia region south of the town—to which, by the way, the Romanian car brand Dacia proudly refers. In the eleventh century, King Ladislaus I of Hungary (1077–1095) founded a bishopric in Oradea and it became an important city with an impressive fortress. The fortress was destroyed and rebuilt several times, getting its pentagonal star-shape between 1569 and 1598. Its equestrian statue from 1390 was the first such public statue in Europe after the Roman period. The city suffered from attacks by the Turks in 1474, 1598, 1658 and 1660, when it finally became part of the Ottoman Empire, until the Habsburgs took it over, in 1692. After the First World War, Oradea became a part of Romania. This was confirmed by the Treaty of Paris in 1947 (Figs. 9.1 and 9.2).

Crucial to the story of the Oradea meridian is that it had an astronomical observatory built in 1459, decades before many other European cities. An influential astronomer, Austrian Georg von Peuerbach (1423–1461), worked for some time at the observatory of Oradea. He used the meridian through the city as a reference meridian in his *Tabula Varadiensis*, published posthumously in 1464. The title of that book, literally *The Tables of Varad*, refers to the meaning of Oradea's Hungarian name of 'Nagy-varad' or 'great castle'. Using this meridian as the prime meridian not only made sense because the observatory was located there, but also it was an acknowledgement to

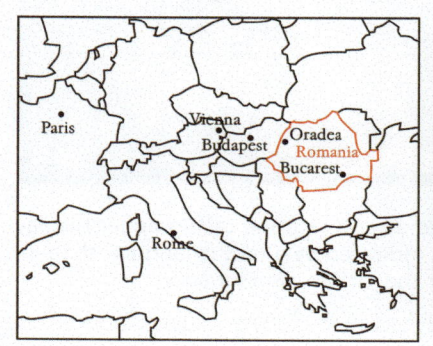

Fig. 9.2 Location of Oradea (left) and an aerial view on the city and its fortress (right)

his main sponsor, the catholic Bishop of Oradea, John Vitez (1408–1472), who had brought him there. Vitez had the support of the Hungarian King Matthias Corvinus (1443–1490) to transform Oradea into the center of modern astronomy at that time.

Von Peuerbach was not just any astronomer. He is sometimes seen as the founder of observational astronomy, paving the road for the Copernican revolution. He earned a reputation through his influential book *Theoricae Novae Planetarum*, a textbook explaining the astronomy of Ptolemy to readers of his day. It became the standard university text on astronomy. Von Peuerbach had studied in Vienna and then traveled all over Europe, from 1448 to 1451. He went to the Italian city-states of Bologna and Padua, where he met Giovanni Bianchini, an astronomer from the University of Ferrara (1410–1469). He returned to Vienna in 1453, where he would remain for most of his time, even after accepting a position at the court of King Ladislas V of Bohemia and Hungary in 1454.

Von Peuerbach had a student who would later become famous, Johannes Müller von Königsberg (1436–1476), better known as Regiomontanus (Regio = king = König and montanus = mountain = Berg). Regiomontanus' tables, based on his work with von Peuerbach, were used by Danish astronomer Tycho Brahe (1546–1601) and later by Johannes Kepler (1571–1630). Kepler and Nicolaus Copernicus (1473–1543) studied Von Peuerbach's *Theoricae Novae Planetarum*. So, it is not surprising that Italian explorer Christopher Columbus (1450 or 1451–1506) owned a copy of the *Tabula Varadiensis*. Even the man who gave America its name, Columbus' compatriot Amerigo Vespucci (1454–1512), remembered how he made calculations based on the *Tabula Varadiensis*. For about two centuries the prime meridian used for navigation would pass through the Oradea.

Oradea Today

The castle with the former observatory was demolished in 1618–1620, by Gabriel Bethlen, Prince of Transylvania (1580–1629). He built the inner castle that can still be seen inside the pentagonal fortress. In 2010–2015 the fortress underwent a restoration process. There is a plate with a plan, but the observatory is not indicated on it. However, when the restoration is finished, the observatory will be indicated by a monument inside the fortress, as close as possible to where it was assumed to be (Fig. 9.3).

In the old city center of Oradea, the interest in astronomy can be seen on its 'Church with the Moon', built between 1784 and 1790. Under the

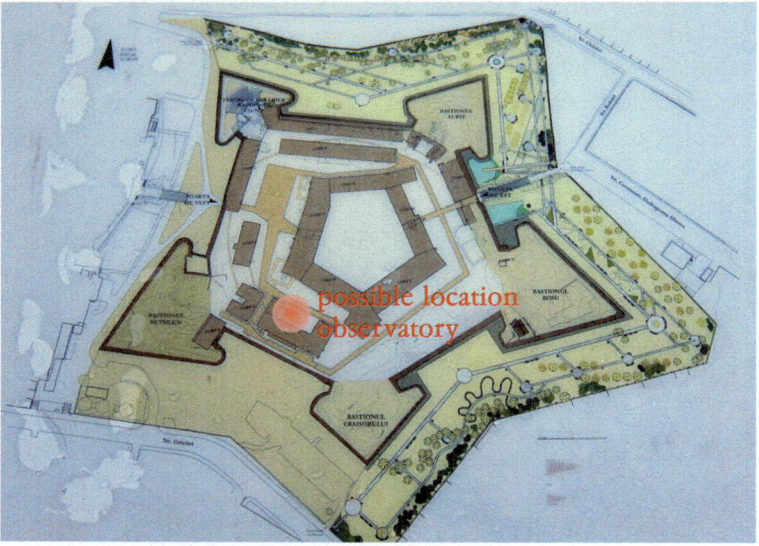

Fig. 9.3 The remains of the once famous Oradea Castle with the possible location of the former observatory

clock tower hangs a sphere showing the phase of the moon on a given day (Fig. 9.4).

Oradea's Astronomical Society is called Astroclub Meridian 0 Oradea and it is the place where astronomy is taught to the town's children. Nicoleta Pazmany and Marin Bica did so for many years, but the latter passed away in

Fig. 9.4 The Oradea Church with the Moon

2013. In memory of the man who helped to restore some of Oradea's former astronomical tradition, asteroid 4633 was named 'Marin Bica', in 2014.

2009 was the International Year of Astronomy and so Romanian nuclear physicist Tibor Toró held a speech for the International Astronomical Union, proposing that the *Tabula Varadiensis* be included in the UNESCO World Heritage List. In September 2016, first Romanian astronaut Dumitru Prunariu held a press conference in Oradea, on the 10th anniversary of the astronomy club).

Each His Own Meridian

Having a meridian is, of course, not a big deal for a city. After all, a meridian runs through every place on earth: it is half of the imaginary great circle on the earth's surface through that place, from the North to South Poles. A meridian connects points of equal length, that is, points that lie as much to the west or east. Today the reference or prime meridian is the meridian passing through the English town of Greenwich. The circle passing through Greenwich divides the world into the western and eastern hemispheres. Thus, Paris, for example, has a longitude of 2° 20′ 56″ E, while Dublin is at 6° 15′ 58″ W (Fig. 9.5).

The prime meridian shifted several times in history, because after all, it can be chosen arbitrarily, unlike an equator which is determined by the Earth's rotational axis. In 220 BC, Greek mathematician Eratosthenes of Cyrene (276–195/194 BC) used his town of Alexandria (Egypt) as a prime meridian, while Hipparchus of Nicaea (190–120 BC) preferred the one through Rhodes. Greco-Roman astronomer Claudius Ptolemy (c.100–c. 170 AD) sometimes used the meridian of Alexandria, where he lived, but more often he chose the left corner of his map: the westernmost point of the then known world. That point corresponded to the Insulae Fortunatae or Blessed Islands, a group of real or imaginary islands in the Atlantic Ocean, which may be associated with the Canary Islands (13°–18° W) or with the Cape Verde Islands (22°–25° W). Because the maps at the end of the sixteenth century, when Ptolemy's maps were reprinted, were not that precise, it was often assumed that the Greek geographer used an island called Ferro (today El Hierro) as the westernmost island, so that the meridian of Ferro became the prime meridian for several centuries (Fig. 9.6).

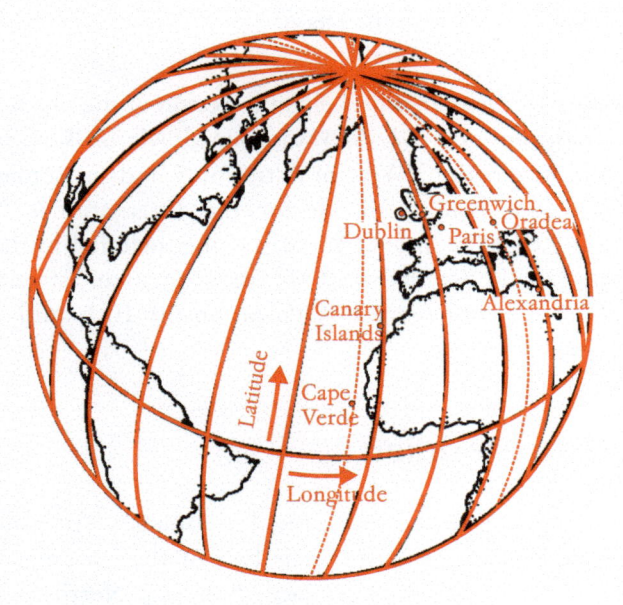

Fig. 9.5 Some of the places mentioned in the text

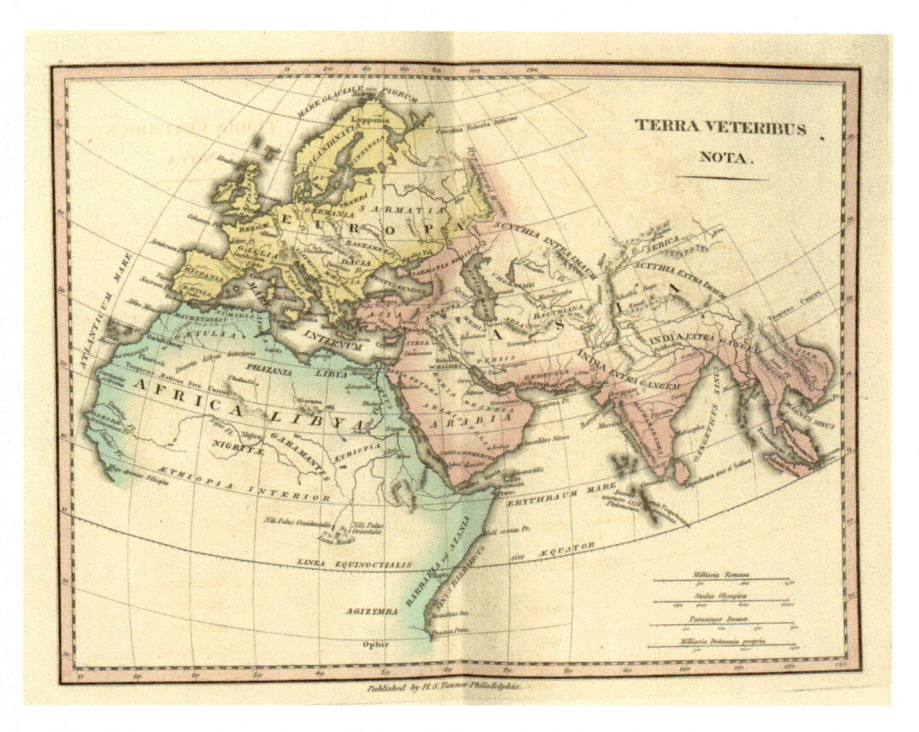

Fig. 9.6 This map from Henry Schenck Tanner's *Atlas of Ancient Geography* was made following Ptolemy and begins on the left with a 0° meridian left of the Cape Verde Islands

On a globe from 1541, Belgian cartographer Gerard Mercator (1512–1594) used a reference meridian running through the Canary Island of Fuerteventura (14° 1' W), but in his *Atlas Cosmographicae* of 1595, the meridian shifts west, to the island of Santa Maria in the Atlantic Ocean (25° W). In contrast to the Antwerp resident, others expressed their chauvinism by invariably situating the prime meridian through their preferred city. The Spaniards took their meridian through Toledo, the Poles through Warsaw, the Danes through Copenhagen, the Swiss through Berne, and the Greeks through Athens. They all had a good reason, such as the presence of an important observatory. And so did the Turks, using the line through Istanbul, the Japanese who preferred Kyoto's, while the Hindus are said to have used the Ujjain meridian, since the fourth century. The Russians thought that the meridian of Saint Petersburg should be the primary one, although that contradicted the choice of the Orthodox Church, which to this day calculates the date of Easter by referring to the meridian of Jerusalem. For a long time, Italians could not agree if Pisa, Genoa, Naples or Rome should serve as the reference. They were defeated in stubbornness though by the French, who held on to their own Paris meridian until 1911 despite an 1884 agreement imposing the Greenwich meridian as the reference meridian for the whole world. After all, Britain ruled the waves—and therefore also the meridians. In addition, British mathematicians had prepared very accurate astronomical tables that were used by most of the sailors. That mathematical reason was precisely why Columbus had chosen the Oradea meridian in his day: through the centuries, the waves were ruled primarily by mathematics.

10

Mathematical Stock Advice

Banking crises make people doubt the reliability of some financial advice. There are good reasons to do so, as some of the mathematical justifications are shaky at best.

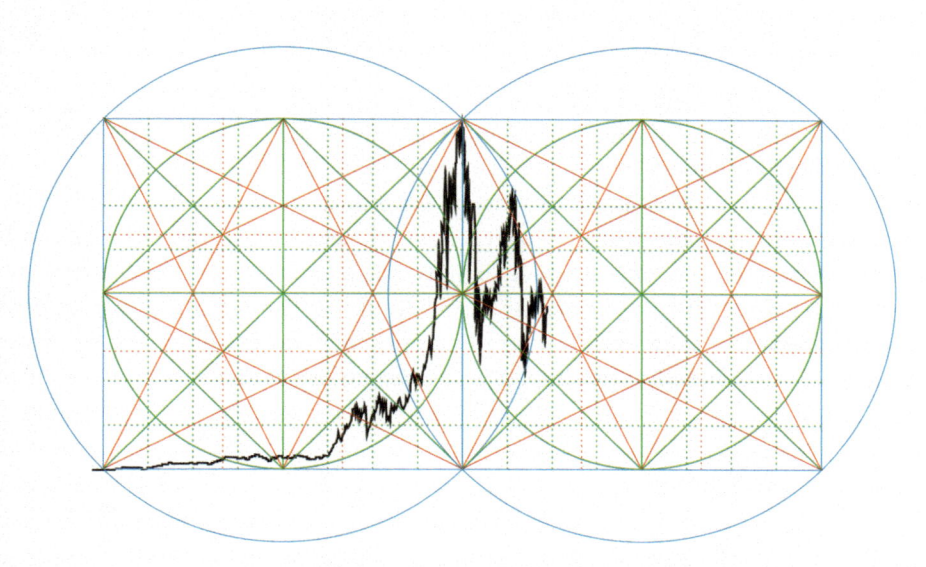

Fig. 10.1 Le Corbusier's geometric mumbo-jumbo reached stock exchange fortune tellers as well

Financial Astrology

It remains popular in art circles to draw all kinds of geometric figures over paintings, sketches or photographs, so that they obtain a 'deeper meaning'. This area of study is somewhat patronizingly labeled as 'holy geometry' and fortunately it does not do much damage. Sure, artists and gallery owners use all kinds of sales pitches to increase the price of an artwork, and so why shouldn't they use mathematical mumbo-jumbo as well (Fig. 10.1)?

This practice of simply drawing geometric shapes over pictures without any statistical justification also exists in stock exchange circles. There is even a subfield of stock market analysis with the telling name 'financial astrology'. In 1998, an article linking stock market depressions in the fall to the phases of the moon won the Charles H. Dow Award given by the Market Technicians Association (MTA) for best technical analysis. In 2014, American economist Robert Novy-Marx published a paper along the same lines entitled 'Predicting anomaly performance with politics, the weather, global warming, sunspots and the stars' in the *Journal of Financial Economics*. The Dutch investment institution Robeco Institutional Asset Management, which holds a license from the Amsterdam Supervisory Authority for the Financial Markets, supported the article in a 2015 blog. In one blog post, Robeco rewrote the entire history of mathematics.

What if people are misled in this way? Can these pseudo-mathematical consultants who use mathematical patterns, about which they fail to understand even the basics, be sued? If an architect without a degree approves a building permit, or a quack performs a medical procedure without being a doctor, he will be sent to court. The nonsense of some stock market drawings can be seen by anyone, so it is not a matter of interpretation or shifting responsibility to someone else. Shouldn't mathematical statements by unskilled advisers be persecuted just as much? A surprising point of view, perhaps, and hopefully this chapter will substantiate this.

Holy Geometry

Together with the number pi, $\pi = 3.14...$, the golden ratio $\phi = 1.618... = (1 + \sqrt{5})/2$ is probably the best-known mathematical constant. It is connected to the Fibonacci sequence: 1, 1, 2, 3, 5, 8, ... in which every number is the sum of the two previous ones: $2 = 1 + 1$, $3 = 1 + 2$, $5 = 2 + 3$, $8 = 3 + 5$, ... and, also in that sequence, the division of consecutive numbers, $3/2 = 1.5$; $5/3 = 1.666...$; $8/5 = 1.6$, approaches

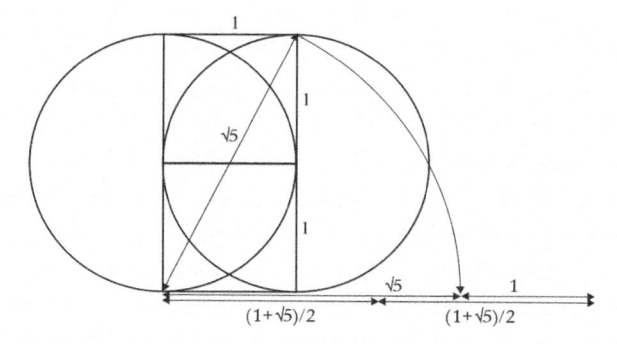

Fig. 10.2 Golden section construction from circles or squares

1.618..., the golden ratio. The number has fun arithmetic and geometric properties because it is the length of the diagonal of a pentagon with side 1. In 1854, however, a seminal publication by German Adolf Zeising spawned a myth attributing exaggerated interpretations to these properties and ascribing mythical qualities to the number. The number ϕ would be the code for beauty, as all kinds of imaginative drawings would prove. The Parthenon is a favorite subject, as are the paintings of Leonardo da Vinci. Scientists look down on these interpretations with suspicion and classify 'drawing polygons on pictures' without the slightest statistical justification in the same category as astrology or homeopathy. Of course, given two intersecting circles or two adjacent squares, the square root of 5 quickly comes up (it is the hypotenuse of a right triangle with sides 1 and 2). Adding 1 and halving the sum is sufficient to obtain the golden ratio, $(1 + \sqrt{5})/2$: in other words, there is nothing to it (Fig. 10.2).

Gradually, the golden section myth also infected the stock market. The latter may be a strange world for mathematicians and for the readers of this book, and so we first provide a short introduction to the financial world.

Stock Markets

Say there is a fruit auction, where apples are bought and sold. The highest bidder goes home with the apples while the seller gets money. On a stock exchange, the act of buying is less tangible because stocks, bonds, options, and futures are traded, which are just 'tradable rights' representing a financial value. When buying a share, the buyer receives a negotiable instrument, as a result of which he becomes economically (co-)owner of a company. If that company makes a profit, the shareholder can get a share of it. That is where the value of the share comes from. A bond is a negotiable debt instrument for a loan provided by a government, corporation, or institution. If a company needs money, it can get financing through a bond. The buyer of the bond receives interest from the issuer. Other papers traded on a stock exchange are options and term accounts.

Regardless of whether the first stock supervisors worked in Mesopotamia, in ancient Rome, in medieval France or in Venice, the fact is that from 1309 onwards, official meetings took place in Bruges in a house of a certain Van der Beurze, for the first trades in security papers. The Bruges concept spread, and not much later people could buy shares in Amsterdam, in particular for the Dutch East India Company. And thus, the Dutch stock exchange is often regarded as the first real stock exchange. Today's most important stock exchange, the one in New York, only took off in 1817. In the past, a stock exchange building was indispensable to keep track of the price of the negotiable instruments, because the price was quoted on a large board. But that has long gone, because today financial values float somewhere in cyberspace, so that a stock market investor no longer actually has a piece of paper in his hand, let alone an apple (Fig. 10.3).

Stock Geometry

Stock prices go up and down, and of course it all comes down to buying something when it's cheap and then selling it when it's expensive. Oddly enough, stock market traders can also 'go short': first they borrow and sell something expensive, and only really buy back it when it is cheap. That defies common sense, and perhaps that situation is unhealthy, but in any case, it means that both rising and falling stock market values have their advantage; a good crisis delights a stock trader. It is therefore important to see trends in stock market charts and try to predict their future values. For example, if a price decreases for a while and if the tops of the temporary peaks are roughly on the same declining line, it is important to know at which point that line is broken. Accordingly, the point where an upward trend is broken is equally important. One can also distinguish zones or regions delimited by lines above and below a graph. A part of a graph then fits, for example, in an ascending or descending triangle (Fig. 10.4).

The Elliott wave principle goes a step further. In the 1930s, American accountant Ralph Nelson Elliott (1871–1948) took pleasure in studying graphs having, for example, three ascending peaks and then two descending ones. A further subdivision yields the Fibonacci numbers: 2, 3, 5, 8, 13,

Fig. 10.3 The building Ter Beurze in Bruges: the first stock market in the world

21, ... After the Polish-French-American mathematician Benoit Mandelbrot gave new life to the concept of a fractal (a geometric figure made up by repeating parts of itself, at infinity, as in a cauliflower) in 1975, some also recognized this mathematical structure in certain stock prices. After all, a stock pattern seems to repeat itself over and over in the same way on a smaller and smaller scale. Today, there are widespread stock exchange sites, supported or not by some psychological theory, claiming people trade in repetitive patterns (Fig. 10.5).

The Fibonacci wave principle may seem questionable, but there are other Fibonacci fantasies. It can, for instance, happen that a price, after it has fallen for some time, moves upwards again for a while, as a kind of upwelling. According to some, this often turns out to be 38.2%, 50% or 61.8% of the previous highest level. Here, they recognize the digits of the inverse of the golden ratio 1.618..., that is, $1/1.618... = 0.618... = 61.8...\%$, those

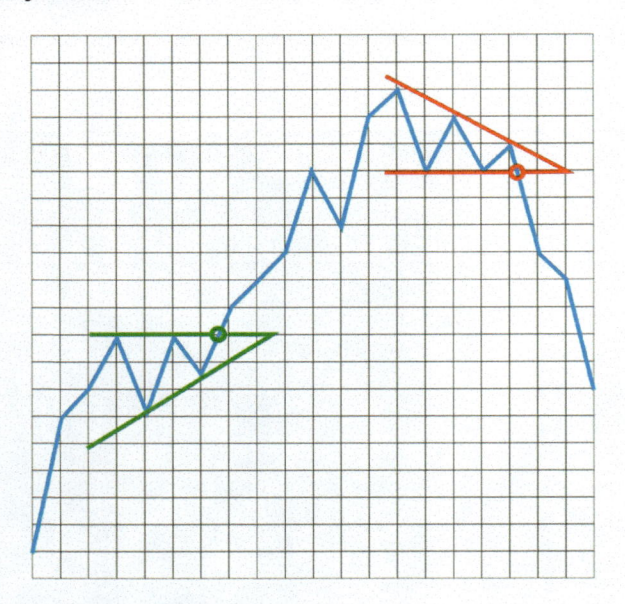

Fig. 10.4 Ascending or descending triangles

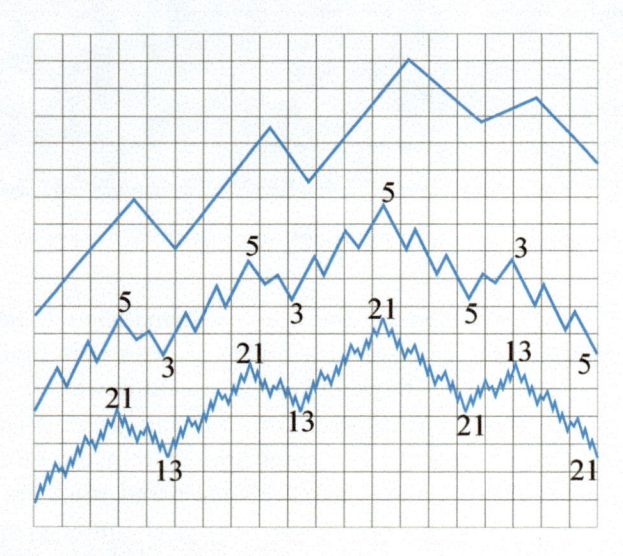

Fig. 10.5 A Fibonacci stock market pattern

of its square, $(0.618\ldots)2 = 38.2\ldots\%$. That so-called 'Fibonacci correction' would also occur in reverse for rising curves that increase to a maximum and then make a short dive downward at 38.2%, 50% or 61.8%. Enthusiasts also discern the cube of $0.618\ldots$, which is $23.6\ldots\%$, and it doesn't stop there.

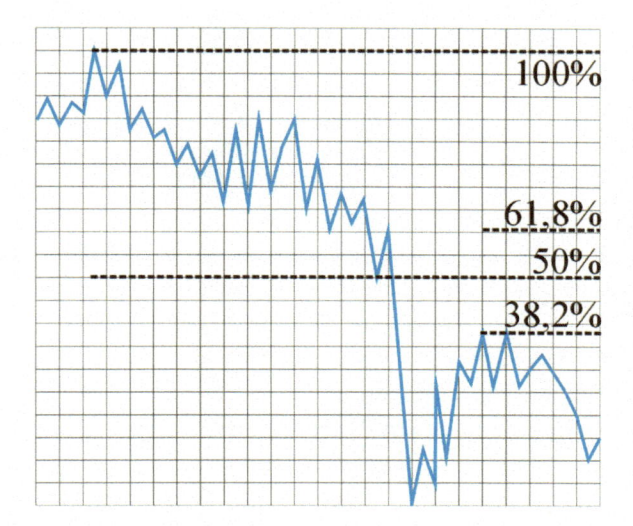

Fig. 10.6 A Fibonacci burp up to 38.2%

Other patterns are also distinguished whether geometric or not. Some drawings involuntarily bring sacred geometric drawings and the related financial astrology (Fig. 10.6).

Fake Predictions

As for financial astrology, there may be some partial truth in it, just as the moon may affect life on Earth as it creates the low and high tides. Though it is another matter to draw far-reaching conclusions from that. Stock prices probably drop in the fall because of the approaching year-end balance sheets and tax settlements rather than because of some moon phase. And the Fibonacci correction can probably be explained without much mystery either. Suppose a price decreases from a certain (temporary) top level to a (temporary) bottom level; set their difference to 1 (or 100%). Suppose half of the investors think the price will rise again to a level x, while the other half estimates on the contrary it will reach a level $1 - x$, then $(½)x = (½)(1 - x)$ or $x = ½$ or 50%. But suppose that the first half thinks that the increase to x% will continue to increase, at x%, then it now follows $x^2 = 1 - x$ and thus $x = 0.618...$ ($x = -1.618...$ is rejected because it is not an increase). Correspondingly, if the mood of the second group prevails, and an increase of x% must be balanced with one of $1 - x$ and another $1 - x$, then $x = (1 - x)^2$ and it follows that $x = 0.381...$ (we reject $x =$

2.618… because that is an increase above the initial level). Sure, these calculations are over-simplified, but in 1994 mathematicians indeed showed how the main patterns arose, through complicated so-called differential equations. University of Pittsburgh professor Gunduz Caginalp co-authored nine publications with Nobel Laureate in Economics (2002) Vernon Smith, and thus these mathematical considerations should not be overlooked lightly. Caginalp disproved a 1973 theory of his compatriot economist Burton Malkiel, who argued that stock prices fluctuate like the path of a drunken man walking along a line (*A Random Walk Down Wall Street*). The latter theory occasionally gets a lot of press attention when a gorilla or an octopus turns out to be as good an investor as, say, Warren Buffet.

The proof of the pudding is in the eating, the reader might think, and so if stock markets are ruled by mathematical patterns, why aren't there some math geniuses who apply their skills to become rich? Well, there are, and the most famous example probable is James Harris Simons (1938). This MIT, Berkeley, Harvard and Stony Brook mathematician worked in one of the most abstract fields of mathematics, while breaking codes for the NSA in between, and setting up a hedge fund company as well, named 'Renaissance Technologies'. Simons' mathematical models and algorithms made him one of richest people in the world, after Forbes (Fig. 10.7).

Fig. 10.7 At the 2014 International Congress of Mathematicians in Seoul, James Harris Simons spoke to over 5000 mathematicians

Still, some argue that long experience on the stock market ultimately plays a bigger role than any mathematical or market theory, and that is a good discussion topic for economists. Here, we only wanted to make the point that mathematics is often visibly misused in stock market advice. And just as astronomers defend themselves against astrologers, mathematicians would do well to stand up for their mathematics. If not, valuable mathematical ideas might get confused with pseudo-scientific ones.

11

The Fall of the Lottery

Many lotteries use rotating drums with numbered balls falling over each other until one of them drops through an opening. They always follow an identical procedure though the time of rotation can differ slightly. The case of the Belgian lottery is used here to find out if this repetitive method really results in a random set of numbers.

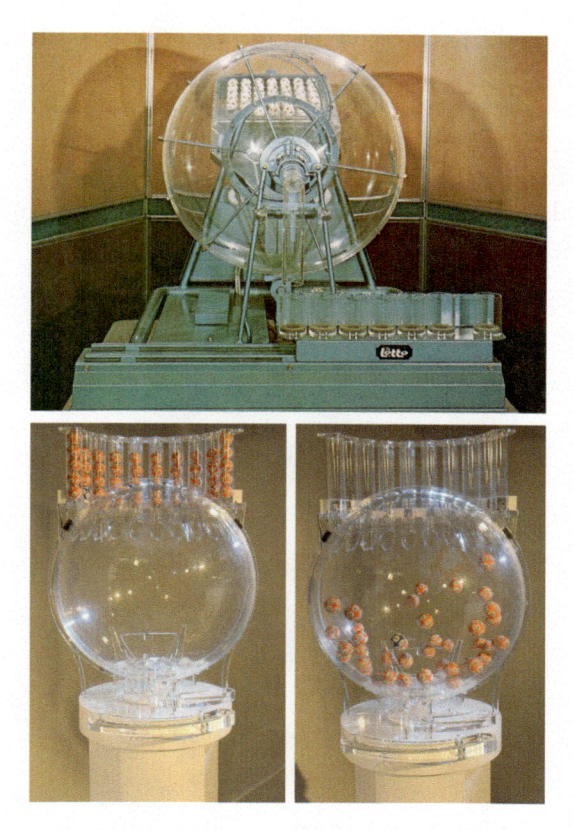

Fig. 11.1 An old lottery drum (above) and a more recent one, before and during the tumbling of the lottery balls

Coincidence or not?

The Belgian National Lottery grew out of the Colonial Lottery, which was officially founded in 1934 to collect money for the Belgian Congo. That message wasn't always shown in an attractive way on the lottery's posters. One poster, for instance, shows white Belgians literally pulling a Congolese person by the hair to get a share of the Congolese fortune. Initially, it only sold the classical lottery tickets with preprinted numbers on a piece of paper, but since 1978 it also organized a weekly lottery where players can choose their own numbers. Between 1978 and 1983, it was seven numbers from 1 to 40, but in 1983 the total was increased to 42, and in October 2011, to 45 (Figs. 11.1 and 11.2).

The chosen numbers are always read on balls that fall into rotating drums. Those drums have been replaced too. Between 1993 and 2005 a drum was used that insiders called 'the diamond', but since 2005 another model was in operation for both 42 and 45 numbers, with ten columns out of which the balls fall into the drum. Yet it is generally known that balls in, for example, a grid structure, do not fall down randomly. If, for instance, there is a triangular pattern preventing the balls from falling straight ahead, they eventually form a so-called 'binomial distribution'. That is, they don't just lie on top of

Fig. 11.2 This advertisement for the Belgian Colonial Lottery invites people to literally take money from the Congolese by playing on the Lottery

each other in a random heap. What about the distribution curve for the balls falling through a lottery mechanism with rotating drums? (Fig. 11.3)

When the drawing of the lottery numbers starts, balls fall from several cylinders almost simultaneously through holes in a drum. That doesn't mean they reach the bottom of the drum in an equal way. Think of people waiting in lines trying to get through a single door: a certain line—be it the left, the right or the middle one—is sometimes squashed by the others. So, it's far from certain that the falling mechanism is as random as it seems. Of course, after that, the rotating movement of the drum makes the balls tumble around in an apparently random way. Therefore, the outcome of a lottery would be unpredictable. But can we imagine that a super powerful computer, programmed according to the laws of physics, could model the dropping and rotation mechanism and that the procedure, therefore, is not as random as it seems? Of course, there is also the innocent hand determining when a ball is selected. But that never takes too long, nor does it happen very quickly. So, if the mechanism and initial conditions are always about the same, wouldn't the statistics rule out the chance factors? Could some law be discerned? We're putting it to the test.

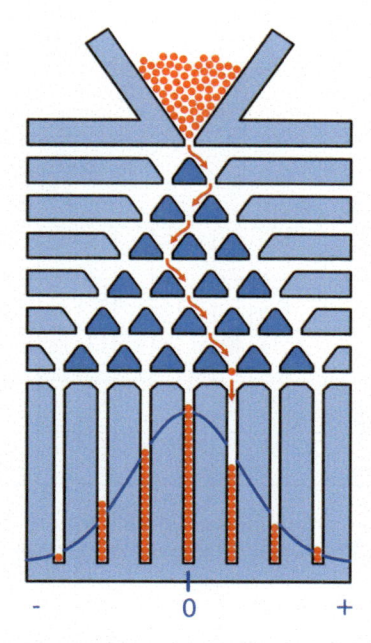

Fig. 11.3 Balls falling on a triangular grid create a binomial distribution. The trajectory of one falling ball is shown

There's Something Going on

The Belgian lottery website contains an overview of all drawn numbers. Despite the fact that the mechanism changed in 1993 and in 2005, all data are lumped together. But there is also a more detailed archive, where a selection by data can be made. Thus, we grouped the results between 2006 and 2011 on the one hand, and those from 2011 onwards on the other hand. Because we do not know when exactly a new drum was put into use in 2005, we use, for safety's sake, only the results between 1 January 2006, and 30 September 2011, and from 1 October 2011, until October 2013 (the time when this text was published in the Dutch science magazine EOS) (Fig. 11.4).

Other tables show the number of consecutive times a number was not drawn on some given day (in this case, the day on which this text was written). In other words, they show the number of weeks since a number was drawn for the last time. The balls drawn in the week the statistic was made carry the number 0. The table is ordered according to the place where the numbered balls are in the starting position of the draw. Below each column is the average of that column. The maxima (red) and the minima (green) seem to indicate that something was wrong with the left side of the mechanism before 2011, while after 2011 the right part shows a surprising disequilibrium (Fig. 11.5).

5									42
4	9	13	17	21	25	29	33	37	41
3	8	12	16	20	24	28	32	36	40
2	7	11	15	19	23	27	31	35	39
1	6	10	14	18	22	26	30	34	38

5	10	15	20	25					
4	9	14	19	24	29	33	37	41	45
3	8	13	18	23	28	32	36	40	44
2	7	12	17	22	27	31	35	39	43
1	6	11	16	21	26	30	34	38	42

Fig. 11.4 Tables with the arrangement of the lottery balls between 2006 and September 2011 (left), and between 1 October 2011, until October 2013 (right)

0									10
2	4	5	0	8	4	15	1	6	0
1	3	11	10	5	4	1	6	12	10
9	2	11	21	3	0	1	4	10	1
0	1	7	1	0	3	0	2	3	7
2.4	2.5	8.5	8	4	2.75	4.25	3.25	7.75	5.6

0	13	5	7	19					
12	1	2	7	1	23	7	1	4	18
0	0	1	2	4	1	2	8	6	5
0	3	11	13	6	8	1	2	6	9
15	3	3	4	3	2	0	7	3	6
5.4	4	4.4	6.6	6.6	8.5	2.5	4.5	4.75	9.5

Fig. 11.5 Tables showing how often a number was not drawn on a given day, giving the average per column in the last rows

Another statistic shows how often a number was drawn in the periods under investigation. We calculated the total per column and compared it in the row below to the expected value per column. Between 2006 and 2011 this follows from the division of the total of draws 4193 (599 drawing days times 7 numbers) divided by 42 balls. Depending on the column, we either multiply the result—99.83—by 5, which is 499.2, or by 4, which is 399.3. After 2011, the total number of drawn balls is 1379 (197 drawing days times 7 numbers). Dividing by 45 makes 30.644, which gives an expected value of 153.22 for columns with 5 balls and a value of 122.58 for columns with 4 balls. Note that both periods are statistically relevant although we have about four times more numbers for the first period than for the second period. Next, we calculated the deviation of the real outcomes from the expected outcome and converted that into a percentage. For both periods, they are surprising for the right-hand side of the table: a difference of −15 and +8% (combined, a 23% gap) for 2011, and a difference of −20 and +15% (no less than 35%!) after 2011. But are these discrepancies sufficient to draw a conclusion? (Fig. 11.6).

	104									93
	92	100	102	97	90	89	99	117	85	97
	106	104	116	103	96	101	107	95	89	98
	105	103	105	87	107	110	119	107	78	101
	97	105	91	92	95	132	94	88	89	108
total per column	504	412	414	379	388	432	419	407	341	497
expected outcome	499.2	399.3	399.3	399.3	399.3	399.3	399.3	399.3	399.3	499.2
relative deviation	1%	3%	4%	-5%	-3%	8%	5%	2%	-15%	0%

	28	28	35	36	27					
	32	25	37	31	31	25	25	28	38	27
	28	37	29	39	28	37	24	37	35	24
	33	30	29	28	33	35	27	32	30	34
	30	25	29	24	39	28	22	23	38	39
total per column	151	145	159	158	158	125	98	120	141	124
expected outcome	153.22	153.22	153.22	153.22	153.22	122.58	122.58	122.58	122.58	122.58
relative deviation	-1%	-5%	4%	3%	3%	2%	-20%	-2%	15%	1%

Fig. 11.6 Tables of the number of times a ball was drawn, and the total per column, the expected outcome and the relative deviation from it

Misled Intuition

For further insight, we consulted David Vyncke of the Department of Applied Mathematics, Computer Science and Statistics at Ghent University, Belgium. His Stochastic Modeling Research Group conducts research in risk management in the financial sector, among others (Fig. 11.7).

> 'It is difficult', he said, 'to draw conclusions from the number of weeks a lottery ball is not drawn. Look at the so-called 'paradox of the longest run'. It states that human intuition has a hard time imagining how many times a certain number can appear or be absent in a series of random numbers. For example, the longest series of non-6 throws in any series of 200 throws with one die is about 22 throws long on average. Intuitively, that seems very surprising, but statistically it is not.'

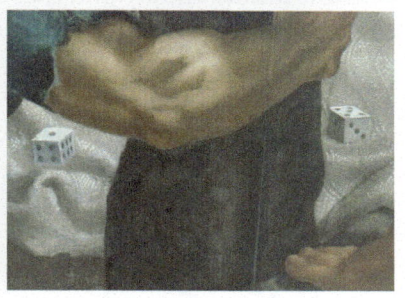

Fig. 11.7 Rolling dice has fascinated people since the beginning of times. In the 1543 painting *Calvary*, Dutchman Maarten van Heemskerck (1498–1574) represented soldiers rolling dice for Jesus' clothes, but one die clearly has incorrect adjacent faces 1 and 6—an indication of their immoral intentions?

Hungarian Tamás Varga set up a seminal experiment to illustrate how easily human intuition can be misled about this. He divided a class into two groups and instructed each student to write a series of two hundred heads or tails on a piece of paper. The students from one group had to flip a coin two hundred times and write down the result, while the students from the other group had to use their imagination and come up with random series themselves. Next, the papers were collected, and Varga tried to match them one by one to the correct group. In most cases he succeeded, because it was enough to count the maximum number of times a student had written heads (or tails) in succession. In a really random series of 200 flips, the longest run is about seven characters long on average, and in 95% of such series, the longest run will be heads at least five times. But a student who makes up such a series by himself rarely writes heads more than four times in a row.

Thus, the number of weeks that the lottery does not draw a certain number can easily give rise to misleading conclusions. The number of times a number occurs could be more informative. Both mechanisms to drop the balls, the one from before 2011 and the one from after 2011, resulted in strange effects on the balls in the columns on the right. Vyncke found the idea of studying the balls per column interesting: the results are striking, at first sight. 'There is a classical statistical method to determine whether a certain deviation is due to chance or not: the chi-square test. That test makes a global comparison of observed frequencies with expected frequencies.' Vyncke performed such a chi-square test on the lottery data. For the situation from 2006 to 2011, where a difference of 23% was observed, he obtained a so-called 'p-value' of 0.10. In the more recent data, where there was a difference of 35%, the p-value was 0.45. This 0.45 deviation is not that exceptional, because it means, simply put, that even if the lottery mechanism was working completely correctly, there is still a 45% chance for such a deviation, and that is fairly acceptable. In the first case, however, the probability is only 10%, and although that is still no reason for a statistician to doubt the randomness of the result, it does raise an eyebrow. However, it is only when the probability is less than 5% that a statistician really rejects the assumption of a perfectly random mechanism.

A Statistical Case?

There is, of course, another factor that influences the draws of the lottery numbers and that is the innocent hand that determines the moment on which one of the tumbling balls is selected. To start off in a simple way, one should perhaps run statistical tests with a stationary drum. The balls then simply fall into the drum from the same starting position. Would there be laws, as with the binomial distribution, if seven balls are always drawn in the same way? There are large sums of money involved in a lottery, so such research could be worthwhile. The difference of 23% for the draws between 2006 and 2011, and 35% from 2011 until now, is certainly striking. The fact that there is only a 10% chance of such anomalies if the Lotto mechanism works perfectly randomly, seems disturbing. However, one cannot turn the argument around and say there's only a 10% chance that the mechanism works correctly. Statistician Vyncke therefore does not draw any far-reaching conclusions:

'There are indeed major differences between the columns, but I think they are due to chance.'

In order not to leave anything to chance, a second opinion was asked, from mathematician Bert Seghers, working for the Royal Flemish Academy of Belgium for Science and the Arts. Like Vyncke, he concluded:

'One cannot claim in 'a statistically ethical way' there is a problem. That, however, does not exclude, nor confirm, that lottery players could benefit from adapting their choice of numbers considering the second last column.'

He added that it would be interesting to check if the strange proportions of -20 and $+15\%$ for the data from 2011 would be confirmed later on. Seghers continued:

'And of course, one could also investigate in a similar way the lotteries of 199 other countries, as they all use similar drawing mechanisms. If all 200 worked perfectly randomly, even then we would expect about 10 lotteries that would (unjustly) not withstand Vyncke's chi-square test, in contrast to the Belgian one, which did withstand the test!'

In any case, references to lotteries being 'a tax on stupid people' are erroneous, as the numbered balls clearly provide food for 'intelligent minds.

12

The Mozarts of Mathematics

In the eighteenth century, Leopold Mozart and his brilliant son Amadeus traveled all over the cultural world of their time. Today, Canadian father Martin Demaine and son Erik amaze the mathematical world, traveling around showing 'mathematics is everywhere, in juggling, in music, in African sand drawings.'

Fig. 12.1 Erik Demaine doesn't want to scare anyone with his math. On the contrary, here he teaches in his Halloween outfit

© The Author(s), under exclusive license to Springer Nature Switzerland AG 2023
D. Huylebrouck, *Dark and Bright Mathematics*, Copernicus Books, https://doi.org/10.1007/978-3-031-36255-2_12

From Greyhound to MIT

The film *Amadeus* showed child prodigy Mozart composing music as a teenager and traveling with his father Leopold all over Europe, from one episcopal court to another imperial palace. Nevertheless, Emperor Joseph II despised 'the many notes' in Mozart's music while the composer's older colleagues, such as Antonio Salieri, looked down with envy on the young, playful genius. At age twenty, Erik Demaine became the youngest professor ever at the prestigious Massachusetts Institute of Technology (MIT), without any regular schooling. He travels around the world of mathematical conferences, together with his father Martin Demaine, 38 years his senior. And like Mozart, Demaine's light-hearted mathematics of origami, magic tricks, unraveling ropes, or dice raises questions among the mathematically unacquainted (Fig. 12.1). An article about this remarkable father and son seemed appropriate, though they explicitly asked for a 'serious paper about their down-to-earth work'.

From age seven, Erik received private lessons from his single father Martin, an artist from Halifax, Canada. When Erik was just six, the puzzle company Erik and Dad Puzzle Co. figured in *The Daily* News of Halifax. Together they traveled the US east coast, while father Demaine was tutoring his son in hotels and on the bus.

> 'But when I had to teach Erik a second language, according to legal regulations for private tutoring, I had a problem, as I only speak English.'

So, he invented a kind of whistling language to fulfill the obligations.

At age twelve, Erik was admitted to the Dalhousie University in Halifax, where he obtained his bachelor's degree at age 14.

> 'We mainly learned together,' Erik says. 'We went to university together, and later it even happened that my father became one of my students.'

Erik surpassed his father, receiving a PhD from the University of Waterloo (Canada) in 2001. It gave him immediate access to MIT (Fig. 12.2).

Erik's publications with titles such as 'C to Java: converting pointers into references' are witness to his initial main interest in pure computer science, but through the 'Computer Science and Artificial Intelligence Laboratory' (CSAIL) at MIT, the young Demaine became interested in mathematics too. He had already shown an interest when he got a hand on the first PCs of the post-IBM era, in the years 1989–1992, 'in the DOS times, remember?'

Fig. 12.2 Erik and Dad Puzzle Co. in *The Daily News* of Halifax

For the then widely used programming language BASIC, some mathematical knowledge was useful.

'That's how I got into mathematics,' says Erik, who started with so-called discrete mathematics and linear algebra. 'By the way', he says, 'a lot of classical calculus is being taught today, but not enough discrete methods, in my opinion.' Calculus is the subfield of mathematics with series, derivatives and integrals, while discrete methods are found in information, set and graph theory or in combinatorics and probability. Mathematically prepared for computer science, Erik then absorbed everything from distributive calculation methods to supercomputing. He learned that programming can be exciting, but also that it has its limits. Ultimately, he focused on algorithms, in which he found a good balance between mathematics and computer science. This science involves the study of finite sets of instructions, usually with some self-repeating steps that cease when a given condition is met.

Magic and Origami

'Algorithms are everywhere', says Erik, 'in juggling, in the geometry of music, in African sand drawings, in the South American knotted quipus'. He acquired the nickname 'the folding mathematician' which oversimplified his achievements, but did give him his international breakthrough. At 17, Erik generalized one of Houdini's magic tricks: any shape bounded by straight lines can be obtained by repeatedly folding one piece of paper and then applying a single scissors cut. For his PhD Demaine solved another folding problem, about planar configurations of rods (not necessarily of equal lengths) connected by hinges, like a carpenter's folding rule. Erik has been asked repeatedly about applications of his works of art (Why is this question never asked about Mozart's work?). There are indeed some very interesting ones, such as the folding of robotic arms, airbags, metal sheets, or even proteins.

'I've been brooding on some problems for ten years,' Erik admits. For example, he remains intrigued by the hundred-year-old question of the hinged conversion of an equilateral triangle into a square (see illustration), with which—rightly or wrongly—the name of Henry Dudeney is associated. 'It seems an arbitrary polygon can always be transformed into another by such hinged intersections,' the Demaines suggest, 'but we have no proof for this, nor is it known what the minimum number of intersections is. Even for Dudeney's solution, it is not certain whether four pieces is indeed the minimum number of puzzle parts' (Note: Demaine partly solved this problem, as discussed in update below.) (Fig. 12.3).

Fig. 12.3 Using hinges, a square can transform into an equilateral triangle

Explanations on Demaine's research are freely available on his website and have found their way into widely available media: there is a documentary film *Between the folds*, and his work on the so-called Mozartkugel—'optimal' folding of a piece of paper around a spherical praline—has been widely commented on. Origami too is taken very seriously by Demaine, and he even learned Japanese to understand it better. Yet, the 'world out there' often has a hard time with his point of view. Even in Japan, 'handcrafted' origami is sometimes looked down upon, and so the terms 'advanced' or 'computational' origami have become more fashionable. Nevertheless, in 2003, at the age of 22, Demaine was awarded a $500,000 MacArthur Fellowship, nicknamed the genius grant. His comment was: 'Winning the MacArthur grant is wonderful because many of the things I do are hard to finance through the regular subsidy channels' (Fig. 12.4).

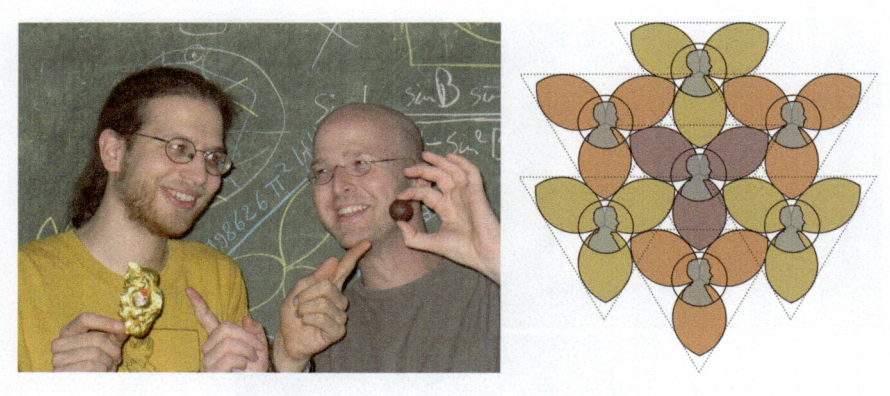

Fig. 12.4 Erik Demaine and Stefan Langerman (ULB, Belgium) showing a Mozartkugel chocolate (left) and their solution for the best wrapping

Playful Mathematics

The Demaines declare their refreshing take on what mathematics is and can be:

> 'Mathematics has become more than advanced algebra. It's a way of looking at things, a way of understanding. And even if you don't understand it, you can do math! Making a work of art or performing about an intriguing problem can allow you to understand a mathematical problem, and perhaps even solve it and thus create new math. We would like to make an interactive sculpture, so that the public can obtain different shapes by folding and hinging the sculpture. That way you can have fun with algorithms, even if the math behind it is difficult to understand. But there are many people who do understand mathematics, even if they don't think so themselves.'

Toys such as Knex or Zometool offer nice challenges and a way to begin thinking about mathematics. At MIT's CSAIL, Demaine artfully plays with a laser cutting steel. 'It is a wonderful material', he thinks, 'because it bends and folds so beautifully. Getting shapes in steel is a new, yet undefined field, because we are only just beginning to understand the mathematics of how a surface bends and folds. Paper can be wrung, but steel cannot.'

Erik Demaine is a 'cooperator': he wants to work with as many people as possible, as is immediately revealed when consulting his CV. His publications rarely say 'I', instead they invariably say 'we'. In the near future, the Demaines intend to write a novella in which ten thousand authors would participate, via an application in artificial intelligence (Fig. 12.5).

Fig. 12.5 Erik Demaine in a light show during a conference in Barbados

Mathematical Improvisation

Mozart was famous for his improvisations. Erik Demaine too prefers teaching 'in real time'. For his university lessons, he dislikes soporific PowerPoint presentations where speakers push the enter key or rattle off memorized mathematical proofs in a kind of playback loop. No, Demaine writes down a theorem and tries to rediscover the proof together with the students. Even if that involves risks, as when he taught the Erdős-Nagy theorem. In 1935 Paul Erdős—known as *The man who loved numbers*, after the book with the same title by Paul Hoffman—formulated a question about so-called 'convex polygons'. These are plane figures formed by straight line segments completely encompassable by a stretched rope. In the case of non-convex polygons there are open spaces left. However, these spaces can always be mirrored around the part of the rope creating the opening, thus providing a new polygon. Erdős suspected that a repetition of this procedure would result in a convex

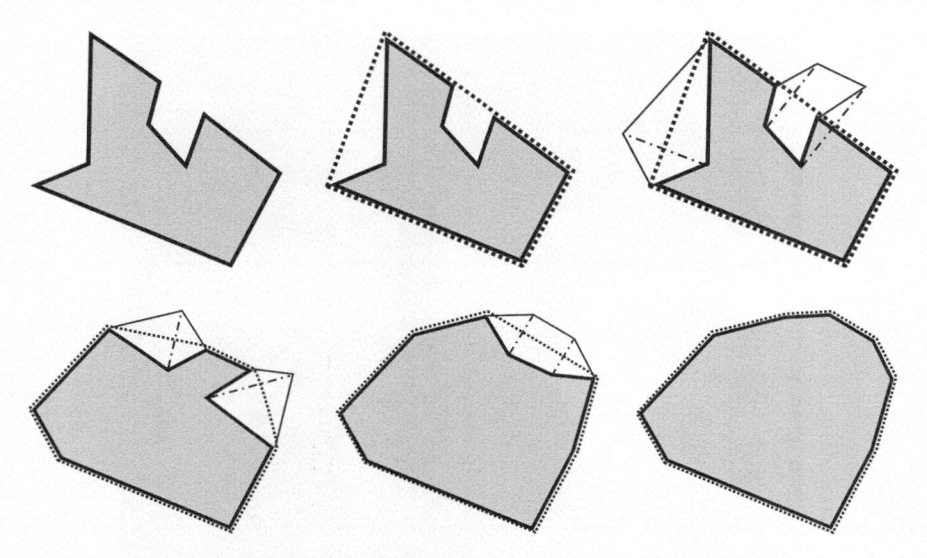

Fig. 12.6 When a rope is stretched around a non-convex polygon (top left) and the openings are mirrored around the part of the rope delimiting them, the (finite) repetition of this procedure yields a convex polygon (bottom right): this is the theorem of 'Erdős-Nagy-Demaine'

polygon, in a finite number of steps. In 1939 Béla Szökefalvi-Nagy thought he had successfully proved this (Fig. 12.6).

It was a beautiful reasoning, which many mathematicians taught and commented on, such as the Canadian Godfried Toussaint and Erik Demaine himself. His MIT students were given a 'modernized' version of Nagy's proof, but one step turned out to be unclear to a student. Demaine couldn't explain it on the blackboard and so he consulted the original 1939 version of the proof. With renewed courage he discussed the issue in the next class, but the student again responded with an 'I don't believe that'. And the student was right; there was a flaw in the proof. Erik Demaine then had to make up a new part for the proof. And yet he doesn't hesitate to repeat the story of the doubting student.

Light-Hearted

Erik Demaine is very fond of his 'Tetris Master' diploma, awarded by the Harvard Tetris Society, on the sixteenth day of the twelfth month of the year 17 Anno Tetri (2002). He was given the prize for proving that Tetris is 'NP-complete', where NP stands for 'Nondeterministic Polynomial-time complete'. Asked for an explanation for a non-mathematically oriented Tetris

Player, Erik Demaine answered, 'Even a computer can't play Tetris perfectly, so humans shouldn't feel bad for not being perfect either'. And for the initiated Tetris players, even the strategy of arranging the blocks as low as possible will not always be optimal. Demaine explains (Fig. 12.7):

'Tetris players should not fear being ridiculed by a computer. But I imagine computers will still outperform humans – it's just saying that neither can be perfect. By the way, chess cannot be played perfectly either by computers, and it is in fact computationally harder than Tetris, though computers now surpass humans – like Kasparov.'

The humor of the Demaines sometimes subtly infiltrates the titles of their publications, as in their article 'Deflating the Pentagon'. Of course, 'pentagon' stands for a five-sided polygon but also for the US Department of Defense, deliberately emphasized by the capital letter. But also more generally, whether it is about mathematics or not, father and son strive for a different way of teaching. When the Demaines first arrived from Canada at MIT, they were disappointed that most students only consulted the Internet and rarely went to the library. Therefore, one of their first assignments was making a library chair out of books only. 'We wanted to up-cycle the books, not just recycle them.' MIT appreciated the interdisciplinary aspect, and the chair was put on display at the Boston Public Library. However, the Demaines don't just literally sit on science, they literally turn it around. Rotations in three dimensions are not necessarily 'commutative' (two successive rotations do not necessarily give the same result if their order is switched) and so they came up with a

Fig. 12.7 A typical Tetris image and Demaine's Tetris diploma

Fig. 12.8 The bicycle helmet is ready on the die, into which Martin Demaine will crawl, pushed around by son Erik, to physically illustrate the mathematical properties of 3D rotations

performance in which Martin crawls into a human size die and is pushed around by Erik (Fig. 12.8).

When invited to give a series of lectures at a rather conservative university in Georgia, they feared a passive and lifeless audience. Erik's first lecture in this series confirmed it, as it got only very polite reactions. Martin then disguised himself and sat incognito among the audience during the next lecture. He asked annoying questions, such as 'Why is your first name, Erik, written with a 'k'?' or 'Do they really pay you to play Tetris?' Halfway through, he stood up showing he could also perform math tricks. When Martin's identity was revealed by Erik, there was relieved applause and a guilty laugh from the audience. But they had achieved their goal: the ice was broken. 'It was one of our best lectures,' they joked.

Update

As the previous account was written in 2009, an update was needed to keep up with the Demaines. For instance, Erik partly proved the above hypotheses that an arbitrary polygon can always be transformed into another by hinged intersections: 'Hinged Dissections Exist', was the very firm title of one of his 2012 papers. Nevertheless, the question about the minimum number of intersections remains open.

The prolificacy of his writings is so huge he surpassed Erdős in number of co-authors. This is quite an achievement, as Hungarian mathematician Paul Erdős (1913–1996) wrote around 1500 mathematical articles, collaborating with 512 co-authors. There even is an 'Erdős number', the minimum number of steps between a mathematician and Erdős in terms of co-authorships. For instance, Demaine's own Erdős number is 2, since he collaborated with Noga Alon, who has Erdős number 1. It was 'confirmed' by his later collaborations with Boris Aronov and with the late Ron Graham who also worked directly with Erdős. But perhaps, as Erdős has passed away, there should be a new reference number, the 'Demaine number', for the minimum number of steps between a mathematician and Demaine.

Curved creases turned out to be one of their most seminal topics, though some hypotheses remain open. Over the years, many a teacher showed how to transform a flat surface into a saddle surface by making concentric circles, but it still isn't certain if it is really possible. The resulting folding would approximate a saddle surface with a negative curvature everywhere, but not the specific shape of a hyperbolic paraboloid, with its typical hyperbolic and parabolic cross-sections, like a Pringle chip or a horse saddle (Fig. 12.9).

And sometimes the results were disappointing: they admit limitations in their understanding of the natural equilibrium paper takes when folded along curved creases, and so they continue to explore this 'self-folding origami' through artwork. Since 2008, they have made curved crease folded sculptures along a given theme. In 2021, for instance, they folded Ludwig van Beethoven's hand-written sheet music of his *Grosse Fuge Opus 133 in Bb major*, while listening to the music, in celebration of the 250th anniversary

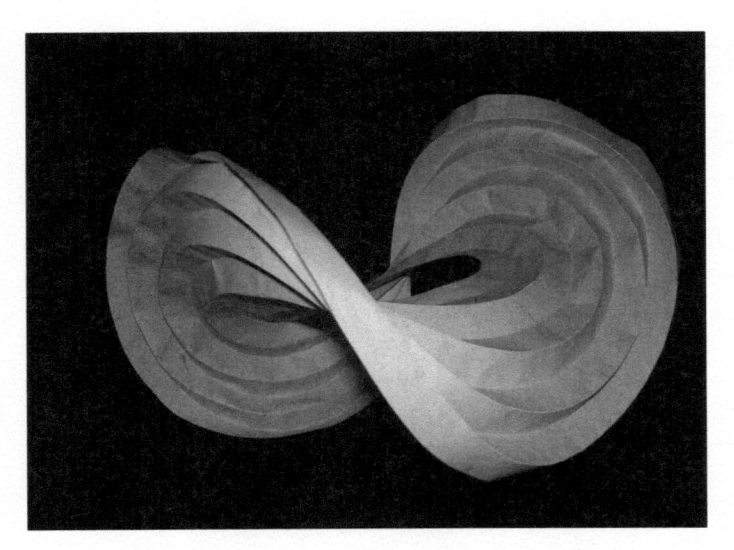

Fig. 12.9 Folding a circular ring into an approximate hyperbolic paraboloid saddle surface

of the composer's birthday the previous year. One of their previous sculptures now belongs to the permanent collection of the MoMA, the Museum of Modern Art in New York (Fig. 12.10).

Creative play is a frequently employed tactic for the Demaines when they think about solving a problem, though games themselves can be research topics too: they wrote papers about Tetris, Mario Bros. and Nintendo. They developed a Tetris font as well, in which all letters are made from falling Tetris blocks. They even designed letter fonts as hinged dissections of a given square, from Sudoku patterns, and so on. They called their collection of fonts 'puzzle fonts', as they explained to a New York Times reporter (Fig. 12.11):

> "To puzzle' means to perplex or confuse, bewilder or bemuse but the word itself is of unknown origin: it's a puzzle where the word 'puzzle' comes from.'

This play with folded sheets of paper and letter fonts has led to some surprising applications. Consider for instance ancient 'locked letters', sheets of paper on which often lengthy letters were written but folded in such a way that they were their own envelope so that no could read the sealed message. Museums don't dare to open them because they are too old or too valuable, but today X-ray microtomography allows their content to be

Fig. 12.10 The Demaines' *Grosse Fugal Form* paper sculpture (2021)

Fig. 12.11 Demaines' 'Tetris Puzzle Font' is made from falling Tetris blocks

reconstituted, using an automated virtual unfolding process the Demaines co-designed (Fig. 12.12).

Another example of applied folding is in making lightweight robots from a sheet of plastic or metal. MIT and Harvard's printable robots project developed techniques for making custom robots out of just $10 or $20 of materials, by constructing a 2D pattern and then folding it into the desired 3D shape.

The seriousness of Demaine's research is illustrated by the many awards he continues to be granted: in 2013, he received the European Association for Theoretical Computer Science (EATCS) Presburger Award for young scientists, as well as a fellowship by the John Simon Guggenheim Memorial Foundation. Two years later, the Nerode Prize followed, and he became a fellow of Association for Computing Machinery (ACM) the year after that. In 2017 an honorary doctorate followed at the Bard College. But perhaps, in the eyes of readers of this book on popular mathematics, the most noticeable award Erik Demaine received, was that he became president of the board of directors of 'Gathering 4 Gardner', a biennial series of conferences in honor of Martin Gardner, the godfather of all popular math book authors.

Fig. 12.12 A simulation of the opening of a letter from 1697. In reality, it was opened only virtually, using computational techniques – and could thus be read (image courtesy Unlocking History research group)

13

Cold and Austere Beauty in Harbin, China

'*Mathematics, rightly viewed, possesses not only truth, but supreme beauty—a beauty cold and austere*" *is a famous statement by British mathematician and winner of the 1950 Nobel Prize in Literature, Bertrand Russell (1872–1970). Beautiful and cold certainly describes the world's biggest ice sculpture festival, in Harbin, China. It mainly features illuminated artistic sculptures up to the size of real palaces, but for the past two winters, the mathematical tourist has also been able to admire a smaller annual exhibition of mathematical ice structures at the Harbin Institute of Technology.*

Fig. 13.1 Peng's surface of revolution on the Harbin Institute of Technology campus (above and middle) and his design for the Harbin Ice Festival

© The Author(s), under exclusive license to Springer Nature
Switzerland AG 2023
D. Huylebrouck, *Dark and Bright Mathematics*, Copernicus Books,
https://doi.org/10.1007/978-3-031-36255-2_13

An Icy Collaboration

Harbin is a town of 8.7 million inhabitants in the north of the People's Republic of China. Because of its Russian past, it has an Orthodox cathedral, a temple that is even more remarkable because it hosts a copy of Leonardo da Vinci's *Last Supper*. Harbin is most famous for its winter ice sculpture festival, attracting up to 300,000 visitors each year. Of course, we often associate paper lanterns with China, but in Harbin, traditional lanterns are also made of ice. Over the years, this practice of carving ice lanterns grew into an ice festival, a kind of frozen Disney World. Giant blocks of ice are cut out of the Songhua River and carved with swing saws, chisels, and ice picks. But the Institute of Technology has been developing new techniques, including one by which reinforced ice is sprayed on balloons, and an even more recent one that pours ice as is done in a rapid prototyping printing machine (Figs. 13.1 and 13.2).

Several years ago, Arno Pronk, of the Eindhoven University of Technology, Netherlands, developed a technique for spraying ice on balloons, inspired by methods for covering curved surfaces with concrete. For greater strength, inspired by the sawdust mixture called Pykrete invented by the Englishman Geoffrey Pyke (1893–1948) during WWII, he added cellulose to water. Pronk often has to explain that this was done as a research exercise with pedagogical advantages, since ice is cheap and reusable, but off the record he honestly confesses that the building of Disney World-like ice structures is his main motivation. This dream resulted in the creation of the world's

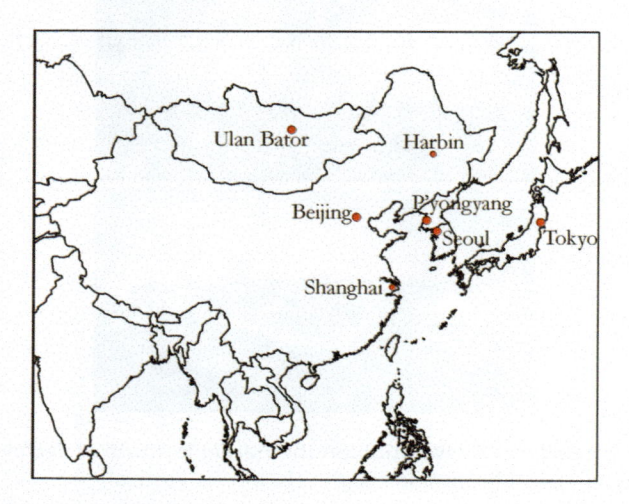

Fig. 13.2 Location of Harbin, P. R. China

Fig. 13.3 Pronk's 2017 pseudospherical construction being made, and its finished version

largest ice dome (2015), a set of towers based on the Sagrada Familia Basilica in Barcelona (2016), and a Leonardo da Vinci bridge (2017), all in Juuka, Finland. In 2018, Pronk moved his operations to Harbin, where there are fewer periods of thaw than in Finland. Also, it plays host to the world-renowned ice festival, and Luo Peng, of the Harbin Institute of Technology, was eager to set up an international collaboration with Pronk. In 2017, this resulted in the creation of a kind of pseudosphere (Fig. 13.3).

In 2018, a larger collaborative project was set up by Peng in which the Harbin Institute of Technology and the Eindhoven University of Technology were joined by Tsinghua University Beijing (China), Kent State University (USA), and the University of Cambridge (UK). The first two had already experimented with ice constructions in Harbin, in 2017, but for Cambridge, it was their first foray into Harbin's frozen world. To involve the Harbin students even more, a series of talks was held, with topics ranging over general information about university collaboration, techniques of ice construction, and mathematical advances in geometry.

Hyperboloids and Cones

John Orr, university lecturer in concrete structures and fellow of Magdalene College, Cambridge, encouraged his students to design and build the world's first hyperboloid in ice. They arrived at their final design through a series of structural stability studies and geometric ideas. A net formed by

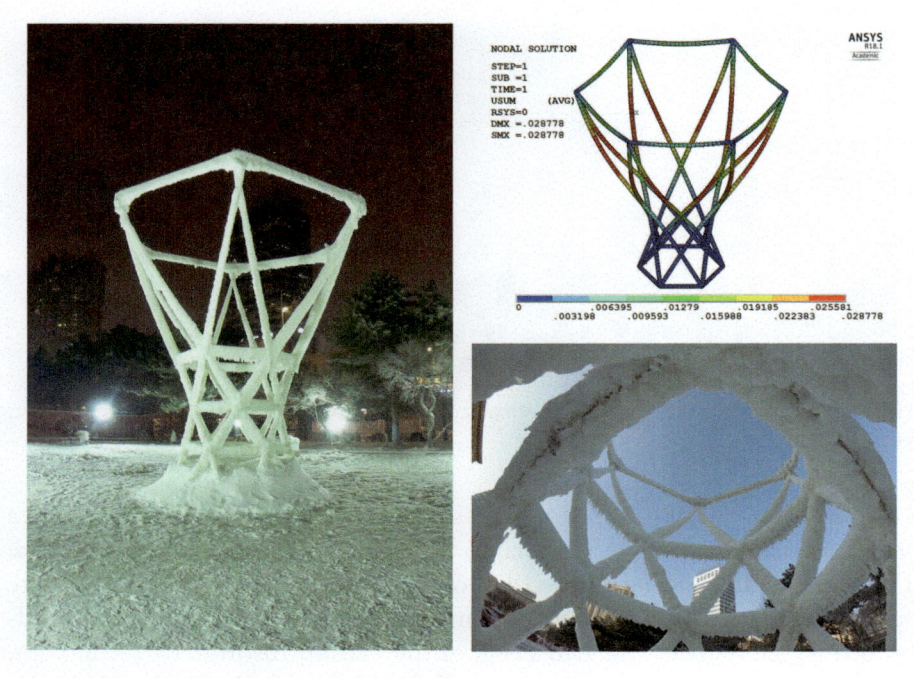

Fig. 13.4 Orr and Millar's student team's hyperboloid

cables and ropes was hung from a hexagonal timber structure and was initially prestressed by the temporary use of a crane. It was sprayed with an ice and cellulose mixture over a period of days to create a structure able to stand by itself. It was all carried out methodically by the student team under the leadership of Cam Millar (Fig. 13.4).

Mark Mistur and Rui Liu, of Kent State University, presented an improved version of their previous year's two cones project, appropriately calling it *Tw-ice*. Ice was sprayed onto two cones with vertical axes formed by fabric hanging on tall wooden pillars. To make them more attractive, the surfaces of the cones weren't perfectly round, but gracefully draped. They intersected along a vertical parabola and had openings at the top that were seen as perfect circles when looked at from a viewpoint along the cones' axes (Fig. 13.5).

Fig. 13.5 Mistur and Liu's parabolic intersection of two cones, with openings on the top that project vertically like circles

Ice Construction Research

In his 2018 project, Pronk used an even newer technique. Instead of spraying, his team of students 'printed' an ice truss dome 1.80 m tall and about four meters wide, mimicking concrete rapid prototyping printing. They did the printing by hand while standing on a scaffolding, pouring ice out of a kind of large, whipped cream bag on ropes stretched over a balloon. Pronk used a special mix of ice, cellulose, guar gum, and xanthan gum to ensure good lubrication of the spray head and better adhesion of the water. He believes that he can omit the rope next time and even automate the process, since it is similar to concrete rapid prototyping printing, in which his university specializes (Fig. 13.6).

For their part, the Chinese teams realized a Möbius strip and part of a surface of revolution. The Möbius strip, with a diameter of 10 m and a height of 3.8 m, was conceived by Professor Weixin Huang, of the School of Architecture at Tsinghua University, in Beijing. Other artworks carrying the name of Möbius, such as the Sarajevo Möbius Bridge, aren't really Möbius strips at

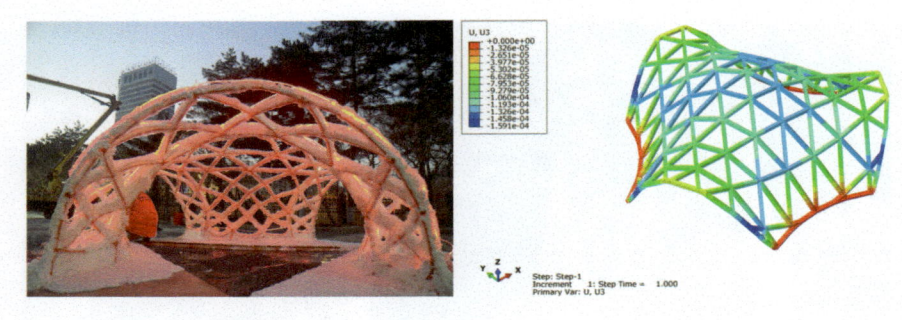

Fig. 13.6 Pronk's 2018 grid shell

all, and only abuse the name, but Professor Huang's Möbius strip was the real thing. Moreover, unlike Max Bill's *Infinity Ring* and other Möbius-themed artworks, Professor Huang carried out structural studies to see whether his construction, based on fiberglass-reinforced plastic rods, would hold up. Professor Huang invented a 'bending active weaving structure system', which can adapt to different freeform geometric constructions. It uses the so-called Kagome lattice, also called a hexadeltille structure, by John Conway. Because of the light effects, the geometric properties of the constituent hexagons and triangles become clearly visible. This made it the world's first structurally thought-through Möbius strip (Fig. 13.7).

Professor Luo Peng, of the organizing Harbin Institute of Technology, made a surface of revolution cut along two planes through the axis of revolution, so that an observer could walk inside and through the icy solid. It was built by spraying water onto a balloon that was removed afterward. Because of his experience in this construction technique, Peng was invited to design a set of seven large halls at the Harbin Ice Festival. This was a first, proving

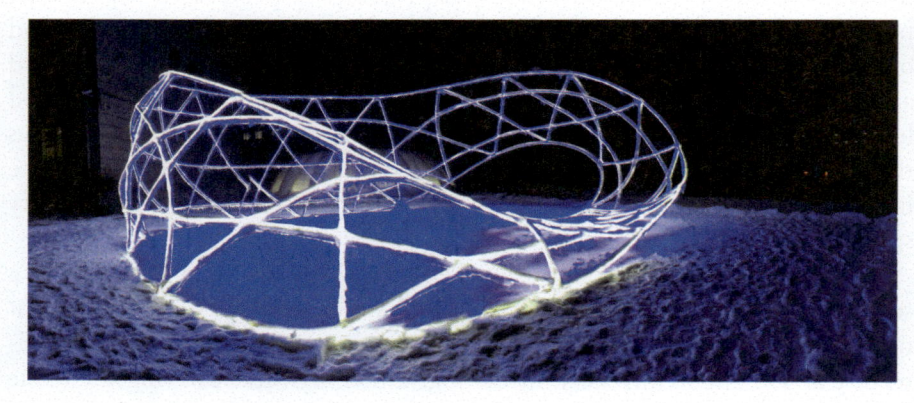

Fig. 13.7 Huang's Möbius ring

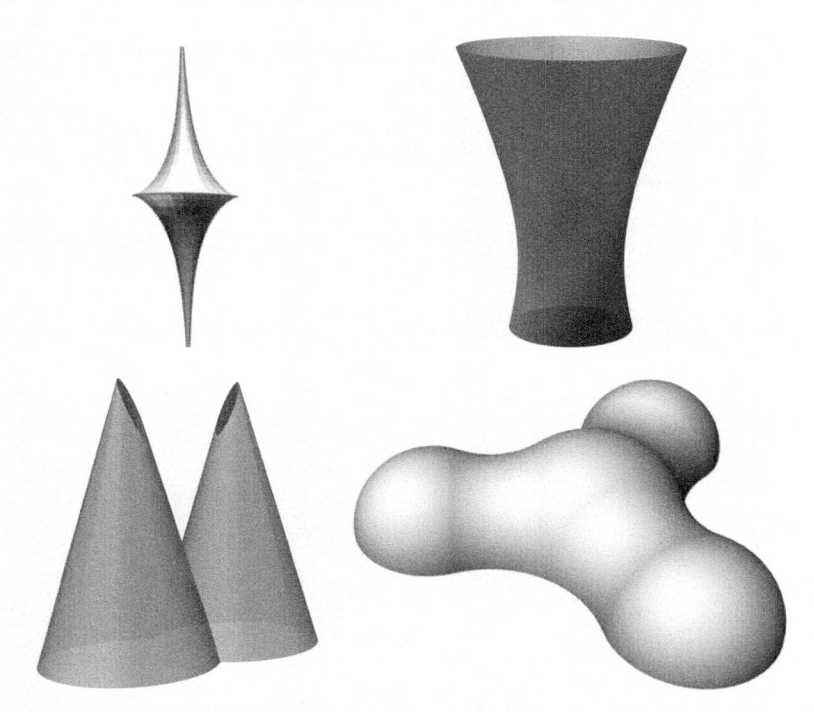

Fig. 13.8 Some of the mathematical shapes mentioned above: the pseudo-sphere (above left), the hyperboloid (above right), two cones intersection along a parabola (below left), and a monkey-saddle type shape formed by 4 half spheres and 3 half hyperboloids (below right)

that structural and mathematical designs continue to find their way to a broad audience. Of course, by the time of publication of this book, the models will have melted, literally. However, the Harbin Ice Festival is an annual event, so one can expect some renewed cold and austere mathematical beauty every year (Fig. 13.8).

14

Your Friends Are More Popular

In the world of social media, most people's friends have more friends than them. However, finding the person with the most friends on, for example, Twitter (X), is a difficult problem. Both questions are interesting mathematical research topics—perhaps too humble for most researchers, but not so for Russian-Dutch Nelly Litvak.

Fig. 14.1 In 2012, a Dutch girl used social media to invite 78 friends to an event, but she had made her invitation 'public', and so thousands of young people showed up. The image on the top shows the distribution of some initial messages and the one below shows the distribution just 22 h later (images courtesy Marijn ten Thij, Indiana University)

© The Author(s), under exclusive license to Springer Nature Switzerland AG 2023
D. Huylebrouck, *Dark and Bright Mathematics*, Copernicus Books,
https://doi.org/10.1007/978-3-031-36255-2_14

The Math of Facebook and Twitter (X)

An important research topic in mathematics is 'networks', like spider webs (real or imaginary), that connect nodes via lines or arcs. These nodes can be computers, where the internet links are the connections, or train stations with railway lines as connections. Less obvious networks could be biological, such as wildlife parks full of animals and plants. The connecting lines are then given by, for example, 'who eats whom'. But of course, the most popular networks these days are those between individuals on Facebook, Twitter (X), Instagram, and the like, connected by relationships between 'friends' or 'followers' (Fig. 14.1).

Mathematicians prefer to speak of a graph, in which lines are connected in nodes. The number of lines or arcs in a node is called its degree. Graph theory has existed for almost 300 years, and all sorts of propositions were developed, and yet in 1991 sociologist Scott L. Feld (1975–1991: State University of New York at Stony Brook; 1991–2004: Louisiana State University; 2004–: Purdue University) came up with a curious conundrum in his article *Why Your Friends Have More Friends Than You Do*:

> 'It is reasonable to suppose that individuals use the number of friends that their friends have as one basis for determining whether they, themselves, have an adequate number of friends. This article shows that, if individuals compare themselves with their friends, it is likely that most of them will feel relatively inadequate'.

In other words, on Facebook or Twitter (X), most people conclude that they have fewer friends or followers than their own friends. It is not a subjective impression, but a fact that can be demonstrated mathematically. Feld drew a diagram for eight friends from Marketville High School. The top left number in each box indicates the number of friends of each girl. It corresponds to the number of lines connecting her to someone else. The number on the bottom right in each box it shows the average number of friends of her friends (Fig. 14.2).

For example, Betty has 1 girlfriend, while her friend Sue has 4 friends. Sue's friends Alice, Dale, Pam, and Betty have 4, 3, 3 and 1 girlfriend(s) respectively and that gives an average of $(4 + 3 + 3 + 1)/4 = 11/4 = 2.75$. When these averages are calculated for each girl, as shown in the drawing, only Sue and Alice turn out to be more popular than their girlfriends. Carol is as popular as her friends, but Betty, Pam, Tina, Dale, and Jane conclude that they are less popular than their friends. So, 6 out of 8 girls, or 75%, find that they are less popular than their own friends.

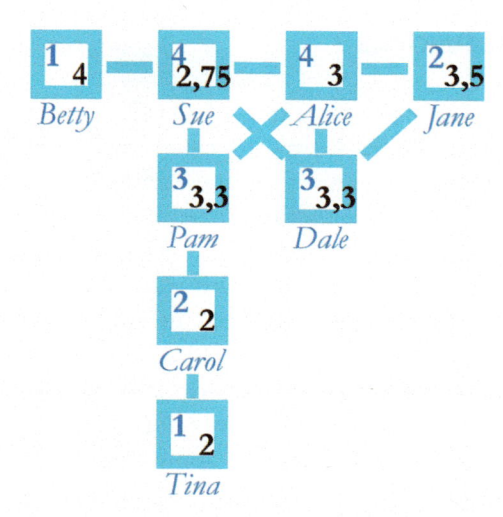

Fig. 14.2 Feld's description of friendships at the Marketville High School

The explanation is that the average number of friends is 2.5, but the average number of friends of friends is 2.98. And 2.5 is of course smaller than 2.98:

- average number of girlfriends = (1 + 4 + 4 + 2 + 3 + 3 + 2 + 1)/8 = 2.5
- average number of girlfriends of girlfriends = (4 + 2.75 + 3 + 3.5 + 3.3 + 3.3 + 2 + 2)/8 = 2.98.

After all, popular people are counted more often, and they increase the average number of friends. Feld's observation may explain why social media makes many people unhappy. That would not only be because success stories are shared more often than failures, but also because the average user concludes that he is less 'popular' than others in his circle of friends. So don't get depressed when you find out that you have fewer Facebook friends than your friends, it's just a matter of mathematical formulas for averages.

Even without mathematics one can intuitively understand why this is so. Imagine a meeting where someone invited nine friends who don't know each other. For example, a politician holding a reception to increase his popularity, or a shopkeeper holding an open house. So, of the ten people at that meeting, one person has nine friends, that is, the politician or the shopkeeper. But there are nine participants with only one friend at that meeting and their only friend has nine friends of his own. Nine out of ten people at that meeting become depressed.

Scale-Free Networks

It was remarkable, and perhaps telling, that this example allowed Russian-Dutch mathematician Nelly Litvak to catch the attention of some 800 teachers during a lecture, organized by the Freudenthal Institute of Utrecht University. Litvak is a top researcher in the field of networks as she is professor of 'algorithms for complex networks' at the University of Twente and at the Eindhoven University of Technology. She studied applied mathematics in Russia, after which she obtained her PhD in applied probability theory at Eindhoven University of Technology in 2002. Her interests lie in the study of large networks, such as the social networks and the World Wide Web, and algorithms that use a degree of randomness (Fig. 14.3).

There is something very remarkable about networks, Litvak explained: the well-known 'Gauss's law', named after the German mathematician Carl Friedrich Gauss (1777–1855), (usually) does not apply. His law is described by a 'bell curve,' or by 'Napoleon's hat curve' and corresponds to the following equation:

$$z = \frac{1}{\sigma\sqrt{2\pi}} e^{-\frac{(x-\mu)^2}{2\sigma^2}}$$

Here π is the famous number pi = 3.14..., the number found when the circumference of a circle is divided by its diameter. The constant e is another

Fig. 14.3 Nelly Litvak explaining to 800 mathematicians why they have so few friends

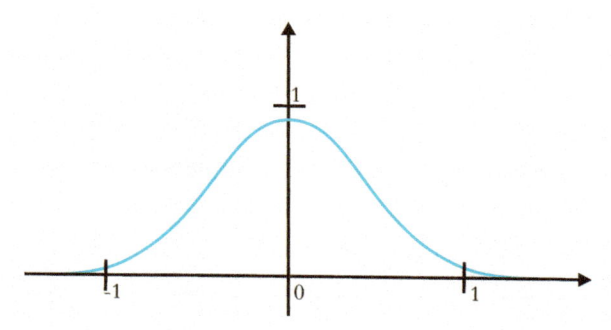

Fig. 14.4 The Gauss bell curve for µ = 0 and σ = 0.44...

famous number, 2.71828..., or Euler's number, after the Swiss mathematician Leonhard Euler (1707–1783). It appears in many mathematical results, such as those related to 'instantaneous rate of interest.' The Greek letter µ (mu) stands for what is commonly called the 'mean,' while σ (sigma) represents the 'standard deviation', indicating how wide the curve (or 'bell') is. In other words, σ indicates how the data is grouped around the mean µ (Fig. 14.4).

The Gauss curve is used in many applications, for instance, when measuring a large number of objects or (body parts of) living beings. For example, measuring the height of 20 year old American men is a Gauss curve with a mean of 5 feet 9 inches (175.4 cm), according to the Centers for Disease Control and Prevention (published in December 2018). But Gauss also pops up when measuring window profiles, or the heights of tree species, or the thickness of bananas, or 'whatever'. But not in networking!

Suppose that p_k is the proportion of nodes with a degree k in a given network. In other words, p_k is the probability that a node in a network has degree k. For example, p_1 is the fraction that indicates how many nodes there are in the network with only one connection, as compared to the total number of nodes; p_2 is this number for the nodes with two connections, etc., so that p_{100} represents the proportion of the number of nodes with one hundred connections, p_{1000} this for a thousand, and so on. In the simple example of the Marketville High School, there were only four of these numbers, p_1, p_2, p_3 and p_4, and they were all equal to $2/8 = 1/4$. In practice, however, as with computer networks, these degrees can be very large.

Well now, says Litvak, most networks do not seem to be governed by Gauss curves, but by power curves. They correspond to the formula:

$$p_k \approx \frac{a}{k^t}$$

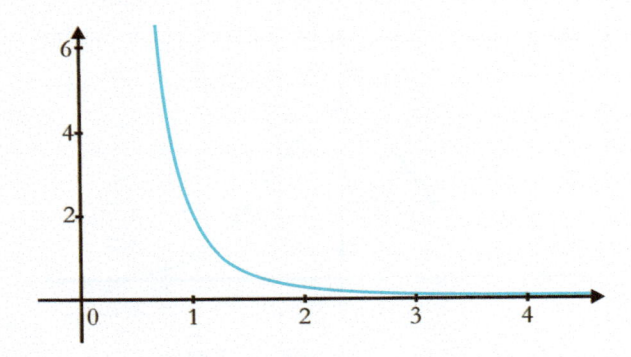

Fig. 14.5 Power curve for a = 2 and t = 3

where t is a number between 2 and 4 and a is a constant (i.e., a given number). The illustration shows the case $t = 3$ and $a = 2$. In applications, the degree k often runs high and therefore a logarithmic scale is used, so that the graph becomes a descending straight line (Fig. 14.5).

Networks whose description fits these power curves are called 'scale-free networks'. They describe the Internet, social networks, biological networks, and so on, although a debate emerged recently on how widespread these 'scale-free networks' really are. However, a paper by Litvak's former Eindhoven colleagues Remco van der Hofstad and Pim van der Hoorn, written together with their actual Northeastern University (Massachusetts) collaborators Ivan Voitalov and Dmitri Krioukov, stated that a good formulation for describing a network is very important. If networks are correctly described, then a power curve formula is indeed the equivalent for networks of the Gauss curve for measured values.

Power of Algorithms

Thus, the approach for networks differs from the ones many are familiar with from statistics. 'New math is needed,' Litvak argues, pointing out that 'a class of optimization problems, which are also relevant for networks, accelerated between 1991 and 2012 as much as 8000-times because of faster computers, and some 469,800 times because of better mathematical algorithms.' Contrary to popular belief, the progress in mathematics has a far greater share in, say, finding the best railway programming, than hardware improvements. The mathematics behind the railway schedules is not yet finished.

Fig. 14.6 Twitter (X) in space: in 2015, ESA proudly announced its Hubble Space Telescope account reached 100,000 followers. *Credit* ESA/Hubble

Another example of a mathematical network problem is the determination of the most followed accounts on Twitter (X). We don't have to explain how important Twitter (X) has become in times when even US presidents communicate through this channel. Of course, you might think, the most popular people would be Obama, Katy Perry, or Justin Bieber, who each share over 100 million followers. But who else? And will they remain popular? Twitter (X) has about one billion accounts and so it would take about 900 years to get all the data for all accounts (Fig. 14.6).

Together with her colleagues, Litvak developed a fast algorithm for such a problem. It relies on the fact that random users often also follow the most popular people (indeed, 'your friends have more friends') and on the fact that the most followed accounts have a lot of followers in a scale-free network (because of that power function). This allowed her to limit the number of accounts to be analyzed. If one searches for the top 10, top 100, or top 1,000 accounts, then she only needs to check $N^{1-\frac{1}{t-1}}$ accounts, if N is the number of users (or nodes). Since t is slightly more or less than 3, this is approximately $N^{1-\frac{1}{3-1}} = N^{1/2} = \sqrt{N}$ accounts. Thus, to find the most popular Twitter (X) accounts, Litvak will certainly not need 900 years. It may be time consuming to learn all of her 'new math' but in the end it will save you ages.

15

The Most Down-To-Earth Problem

If a number is even, divide it by 2. If it is odd, multiply it by 3 and add 1. Repeat those steps. It always ends up with 1 and no one knows why. Mathematicians, however, agree on one thing: recently, world-renowned Terence Tao made strong progress.

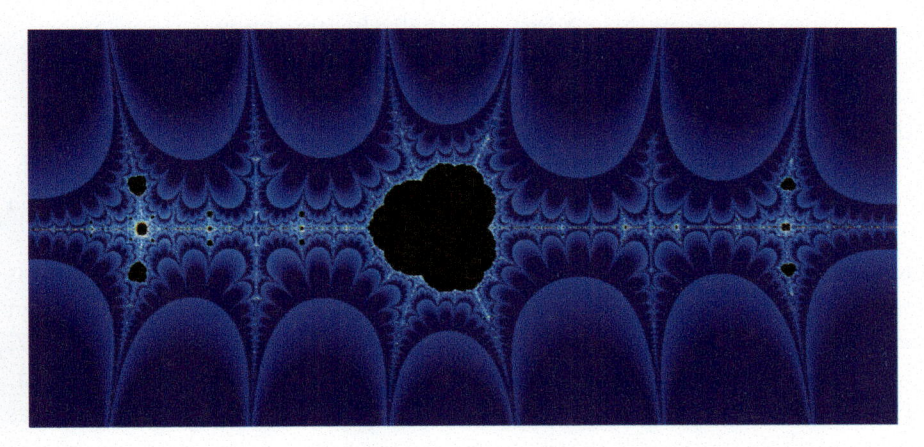

Fig. 15.1 Representation by Nathaniel Johnston (Mount Allison University, Canada) of the Collatz function $f(x) = (x/2)(1 + (-1)^x)/2 + (3x + 1)(1-(-1)^x)/2$. We recognize '$x/2$' and '$3x + 1$'

Mathematics with a Warning

For about thirty years it haunted me: the Collatz conjecture. I still remember my first encounter with it. I was teaching in Burundi and accompanied guest professor Ivan Cnop (University of Brussels) on a 'safari'. I thought I would please the liberal mathematician by saying: 'You see, for animals it is simple: eat or be eaten. That is the meaning of their life, they don't need a god.' To which he replied:

> 'Simple things can sometimes be mysterious. Take 'Collatz's conjecture.' If a number is even, divide it by 2, if not, multiply it by 3 and then add 1. In the long run, the final result is always 1, but no one knows why.'

It does indeed seem a very simple problem. Start with, say, 7. It's an odd number, so we calculate $7 \times 3 + 1 = 22$. That's even, so 11 becomes our next number. Next, we do $11 \times 3 + 1 = 34$, then 17, then 52, 26, 13, 40, 20, 10, 5, 16, 8, 4, 2 and finally we get 1. We could continue and do $1 \times 3 + 1 = 4$, then get 2, and so end up again with 1, as we get in a loop: $1 \to 4 \to 2 \to 1 \to 4 \to 2 \to 1\ldots$. German mathematician Lothar Collatz (1910–1990) conceived the problem in 1928 and presented it as a challenge during his lectures. It gained such popularity when he was talking at Syracuse University (New York), that it also became known as the 'Syracuse conjecture'. Still, it also carries the names Ulam conjecture, Czech conjecture, Thwaite's conjecture, Hasse's algorithm or Kakutani's question. Those different names probably illustrate its esteem.

The conjecture is true for any numerical example one can come up with. It was checked for all numbers less than about 300 trillion, a trillion being a 1 followed by 18 zeros, or a billion times a billion. For some large numbers, testing it is simple, such as for powers of 2, as only consecutive halving is required $(\ldots \to 128 \to 64 \to 8 \to 4 \to 2 \to 1)$. But other numbers, even very small ones, sometimes require very long calculations. The number 27, for example, needs 111 steps. Yet, Collatz's assumption does remain correct. However, all these numerical verifications do not constitute proof. That has upset mathematicians for nearly a hundred years now (Figs. 15.1 and 15.2).

To see where the difficulties lie, let's consider two similar yet different cases. First, suppose we still divide by 2 when a number is even, but just add 1 when it's odd, without multiplying by 3. For the number 7, that we considered above in the case of the original Collatz conjecture, we now obtain $7 \to 7 + 1 = 8 \to 4 \to 2 \to 1$. That's only four steps instead of sixteen. For 27 it goes much faster too: $27 \to 28 \to 14 \to 7 \to 8 \to 4 \to 2 \to 1$. Here too, there is a loop when continuing, although it is a shorter one: $1 \to 2 \to 1$

$97 \to 292 \to 146 \to 73 \to 220 \to 110 \to 55 \to 166 \to 83 \to 250 \to$
$125 \to 376 \to 188 \to 94 \to 47 \to 142 \to 71 \to 214 \to 107 \to 322 \to$
$161 \to 484 \to 242 \to 121 \to 364 \to 182 \to 91 \to 274 \to 137 \to 412 \to$
$206 \to 103 \to 310 \to 155 \to 466 \to 233 \to 700 \to 350 \to 175 \to 526$
$\to 263 \to 790 \to 395 \to 1186 \to 593 \to 1780 \to 890 \to 445 \to 1336$
$\to 668 \to 334 \to 167 \to 502 \to 251 \to 754 \to 377 \to 1132 \to 566 \to$
$283 \to 850 \to 425 \to 1276 \to 638 \to 319 \to 958 \to 479 \to 1438 \to$
$719 \to 2158 \to 1079 \to 3238 \to 1619 \to 4858 \to 2429 \to 7288 \to$
$3644 \to 1822 \to 911 \to 2734 \to 1367 \to 4102 \to 2051 \to 6154 \to$
$3077 \to 9232 \to 4616 \to 2308 \to 1154 \to 577 \to 1732 \to 866 \to 433$
$\to 1300 \to 650 \to 325 \to 976 \to 488 \to 244 \to 122 \to 61 \to 184 \to 92$
$\to 46 \to 23 \to 70 \to 35 \to 106 \to 53 \to 160 \to 80 \to 40 \to 20 \to 10$
$\to 5 \to 16 \to 8 \to 4 \to 2 \to 1$

Fig. 15.2 The number below 100 that requires the most steps is 97: only after 118 steps does the calculation stop

$\to 2 \to 1\ldots$. This method apparently also always yields 1 in the end, but in this case, it can be proven. Half of an even number is always smaller than the number itself. For an odd number, '1 more than that number' is of course greater than the original number, but it is even, so that the next number is half of it, and it is half of the original number plus a half. That's smaller than the original number—as long as it's greater than 1. And so, we always get a smaller number, either on the next step, when it's even, or two steps later, when it's odd. When 1 is reached, the argument doesn't work anymore, but then the end point, 1, has been reached.

In the case of the original Collatz conjecture, that reasoning does not work. If we multiply an odd number by 3 and then add 1, it will be even. But the number that follows, half of that sum, is then 1.5 times the original number plus 0.5. Now, we cannot conclude that it is smaller than the original, on the contrary. So, we now get the impression that the numbers would get bigger and bigger, because of that multiplication by 1.5.

Vain Attempts

It should be noted, and it has been checked, that if one starts with $2^{100}-1$, a huge number, then after about two hundred steps the even bigger number $3^{100}-1$ turns up, which is about a billion times a billion times larger. Sure, with every even-numbered step, there's a division by 2, which reduces the result, but why would this always compensate for the multiplication by 3 and the addition of 1?

Fig. 15.3 The protagonists of this chapter: Lothar Collatz (left) and Terence Tao (right)

Suppose, for instance, that instead of multiplying by 3 and then adding 1, the number 1 is subtracted after the multiplication by 3. For the number 7, which was considered in the two previous examples, something remarkable occurs: now 1 is never reached! Indeed: $7 \to 7 \times 3 - 1 = 20 \to 10 \to 5 \to 14 \to 7$ and that is the starting point. The number 27 also gets in the loop of 7, without reaching 1: $27 \to 80 \to 40 \to 20 \to 10 \to 5 \to 14 \to 7\ldots$. And there are even more difficulties: the following numbers 'turn around' too, without reaching 1: $17 \to 50 \to 25 \to 74 \to 37 \to 110 \to 55 \to 164 \to 82 \to 41 \to 122 \to 61 \to 182 \to 91 \to 272 \to 136 \to 68 \to 34 \to 17 \to \ldots$. Moreover, there is the same loop as in the previous case: $1 \to 2 \to 1 \to 2 \to 1\ldots$. (Fig. 15.3).

Thus, by simply adjusting 'times 3 plus 1' to 'times 3 minus 1', the situation changes completely. So, proving Collatz's conjecture is not as obvious as the numerical examples make it seem. Still, there were a lot of fruitless attempts: Jeffrey Lagarias (University of Michigan) collected them in his 2010 book *The Ultimate Challenge: The 3x + 1 Problem*. And so, just about every popular science article about Collatz's conjecture comes with a warning: don't try this at home, stay away from it, don't waste your time with it. In an article in *Quanta Magazine*, Lagarias stated: 'This is a dangerous issue. People get obsessed with it and it's really impossible.'

Tao Tried His Luck

Yet, Terence Chi-Shen Tao (1975–), an Australian-American mathematician who currently works at the University of California in Los Angeles, sees Collatz's calculation game as a good intellectual challenge. To him, it offers a non-obvious model of a 'dynamical system', where a function describes a point in space depending on time, such as a planet's position, or a moving pendulum. So-called systems of (partial) differential equations, with which every first-year student in a scientific field is familiar, are preferred tools in studying those dynamical systems. Some initial conditions sometimes return finite values after calculation, and sometimes they don't. A typical example are the equations that describe the weather. Under certain initial conditions, the weather evolves gradually, under other conditions a hurricane develops.

'Moreover', says Tao, 'Collatz' formulation is very simple, so that it can serve as a toy model for our mathematical knowledge'. On 8 September 2019, he posted proof for Collatz's conjecture, for 'almost' all numbers. He is one of the most important mathematicians of our time and so his work could not be consigned so easily to the list of aimless attempts. And in the end, mathematicians agreed: while it is not complete proof of the conjecture, it is a very big step forward. Tao proved that the numbers in the successive steps of the Collatz process eventually become as small as desired, for 'almost' every number. Incidentally, the core of Tao's proof was particularly inspired by a dynamical system that Belgian Princeton IAS mathematician Jean Bourgain (1954–2019) had studied in 1994, namely the system that corresponded to the so-called non-linear Schrödinger equation.

Tao realized he could do something similar for Collatz's conjecture: start with a sample of certain initial numbers. If in the Collatz calculations almost all numbers eventually yield 1, then he could conclude that Collatz's conjecture is true for almost all numbers. Of course, it is not that simple: the choice of a good 'sample' is very complicated. Yet, Tao was able to conclude that 99 percent of starting values greater than 1 quadrillion (one million billion) eventually tend to values less than 200. He remained very humble about his work: 'I was able to get so close to the Collatz conjecture if I wanted to, but it was still out of my reach.'

One can only try to guess why Tao took the risk of tackling Collatz. Perhaps the answer lies in his past: from 1992 to 1996, he was a student at the University of Princeton, of Elias Stein (1931–2018), an Ashkenazi whose parents had fled to the US in 1940. Later, he worked closely with Jean Bourgain, with whom he jointly won the prestigious Crafoord Prize for Mathematics in 2012. When Tao learned of Jean Bourgain's death in

December 2019, almost exactly one year after Stein's, he wrote: 'Jean Bourgain and Eli Stein were the two mathematicians who most influenced my early career. It's a shock to find that they're both gone now.' And so, perhaps, that's why he focused on a problem that would immortalize its solver. Tao had already won the Fields Medal in 2006, the most important award for mathematicians under the age of 40, and he usually doesn't deal with 'impossible problems.' Except for a few days a year when he allows himself to 'try his luck', with matters like $\div 2$ or $\times 3 + 1$.

Second Opinion

Philosopher Jean Paul Van Bendegem (Brussels, Belgium) was kind enough to provide a second opinion on what makes Collatz's conjecture so fascinating. He had explained it in his article 'The Collatz Conjecture: A Case Study in Mathematical Problem Solving'. The article illustrated the mathematical-logical techniques involved in the search for a proof. To him, these steps are:

(1) Test some specific cases to get ideas for the general case.
(2) Try out graphics to see if there are patterns to be found.
(3) Come up with the 'right' concepts that can help find evidence because they provide insights.
(4) Devise arguments pointing towards a proof in which mathematical intuition plays an important role.
(5) Associate it with other proven mathematical theorems in the hope that elements in those proofs may help (Fig. 15.4).

'The search for a proof is not a blind wandering', Van Bendegem says, 'and Collatz's conjecture is an important case study for understanding mathematical practice. Especially since the proof is still lacking.' And for readers who may not have had enough of it, Van Bendegem also added a challenge in line with Collatz's conjecture: why can't a chessboard, of which the two diagonally opposite corner squares are removed, be covered with dominoes? Again, a clear and simple problem, which also concerns theorists in the field of mathematical proof.

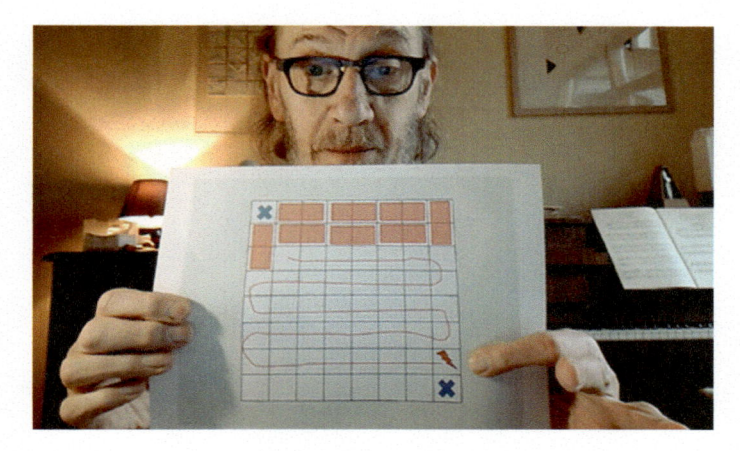

Fig. 15.4 Jean Paul Van Bendegem presenting another Collatz-type problem: why can't a chessboard, of which the two diagonally opposite corner squares are removed, be covered with dominoes?

16

Meccano Math

Mathematics with Meccano sounds like a fun, constructive activity. A Nobel Prize winner in physics, a math education professor, and a research mathematician agree. One can even tackle age-old issues that are unsolvable according to classical drawing methods. And isn't hands-on math ideal in times where technicians are needed?

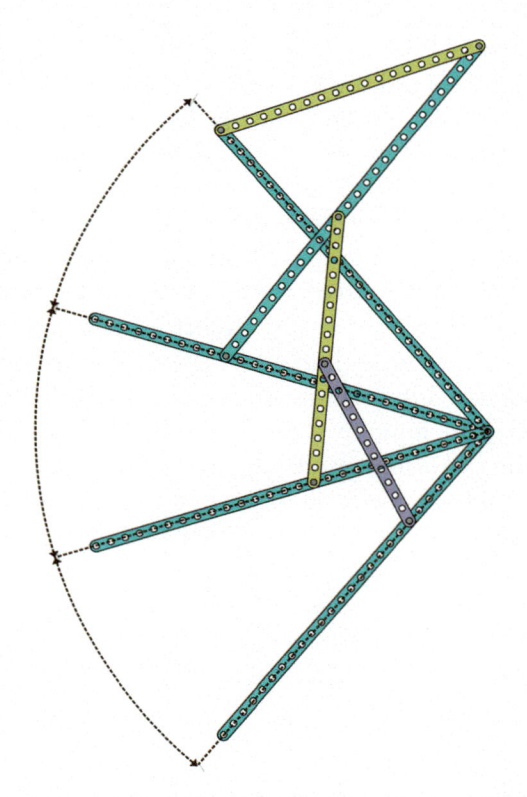

Fig. 16.1 A Meccano construction for the trisection of an angle (dotted lines)

© The Author(s), under exclusive license to Springer Nature
Switzerland AG 2023
D. Huylebrouck, *Dark and Bright Mathematics*, Copernicus Books,
https://doi.org/10.1007/978-3-031-36255-2_16

Nobel Meccano

Mathematics education professor Bart Windels (University of Brussels) asked me to review a paper about *Meccano mathematics*. As editor-in-chief of the Flemish journal for mathematics teachers, he shrugged off his usual neutrality, adding the comment that he found this a very interesting topic. Moreover, it was written by research mathematician Matthias Floré (Mathematics and Data Science, University of Brussels), who was so fascinated by this Meccano mathematics that he programmed a construction game with the mathematical drawing software TikZ. And who has never played with Meccano—or with its plastic variant LEGO®Technic (see Fig. 16.1)?

Apparently, Dutch Nobel Prize winner in physics Gerardus 't Hooft was also fascinated, because the submitted article was based on his work. He wrote three articles on 'Meccano Math', in 2006, 2008 and 2014, and thus it can be said that he is something of an expert. 't Hooft (born in 1946) won the Physics Nobel Prize in 1999, together with his thesis advisor Martin Veltman (1931–2021). He always remained a professor at the Spinoza Institute of Utrecht University. Among his most famous PhD students is Robbert Dijkgraaf, director of the Princeton Institute for Advanced Study from 2012 to 2022.

't Hooft does not shy away from debate. Like Einstein, he is not a big fan of indeterminism or randomness in quantum mechanics, although it is precisely in this field that he received his Nobel Prize. He has his reservations about string theory as a mathematical explanation for the universe, although he teaches that theory himself. He disagreed with Stephen Hawking when he claimed that information is lost in a black hole—and Hawking ultimately agreed with him. In 2008 he objected to an already approved PhD defense stating anti-matter falls upward and rejected the promotion of author Marcoen Cabbolet, twice, at different universities in The Netherlands. The latter became a scientific asylum seeker in Belgium and promoted in 2011 with the same formerly rejected thesis under philosopher Jean Paul Van Bendegem (University of Brussels). However, according to a 2022 CERN experiment, antimatter does fall downward, as 't Hooft predicted. Perhaps all those exhausting discussions are the reason for his interest in the innocent Meccano game? (Fig. 16.2).

Fig. 16.2 Gerard 't Hooft

Meccano Polygons

't Hooft describes how he explored the possibilities of Meccano mathematics. For the construction of a regular pentagon, for example, he proposes different constructions. His first solution uses 11 strips, 6 with 3 holes and 2 with 4, and 1 with 5, 6, and 7 holes each. It translates almost literally the Pythagorean theorem and the property that the diagonal of a regular pentagon with side 2 is equal to $1+\sqrt{5}$. His more elegant solution uses only 9 strips, 6 with 13 holes, 2 with 12 holes and 1 with 10, where the strips overlap each other less. We illustrated the first case with some calculations, but not the second, as this is not necessary with such a 'nice mathematical construction.

Moreover, there is a purely scientific reason for the enthusiasm of a Nobel Prize winner and a mathematics education professor. The latter appreciated how 't Hooft was able to illustrate certain major problems in the history of mathematics with Meccano. For example, there is the construction of the heptagon. Unlike the pentagon, it cannot be drawn with a compass and ruler in a finite number of steps. This also applies to regular polygons with nine, eleven, and thirteen vertices, but then again not for the heptadecagon or 17-gon, which can be drawn that way. Good approximate solutions have been proposed in many cultures, as can be seen in the many near-regular polygons in and on mosques (see examples given in this book). Now one can propose very elegant solutions with Meccano, without encountering the compass-and-ruler limitation (Figs. 16.3 and 16.4).

The problem of the construction of polygons is related to the three great problems left by the ancient Greeks and on which all mathematicians would later ponder:

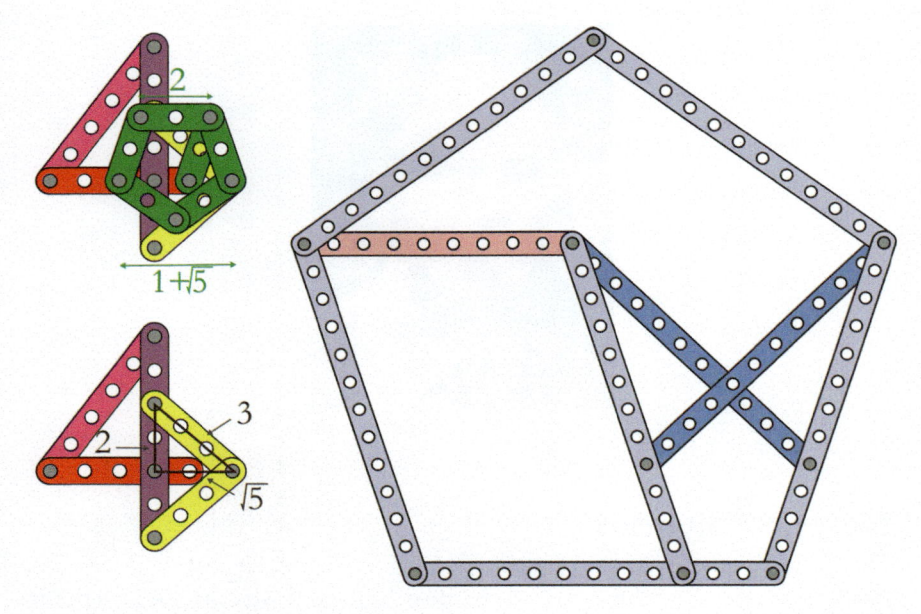

Fig. 16.3 Mathematics with Meccano: two ways to construct a regular pentagon

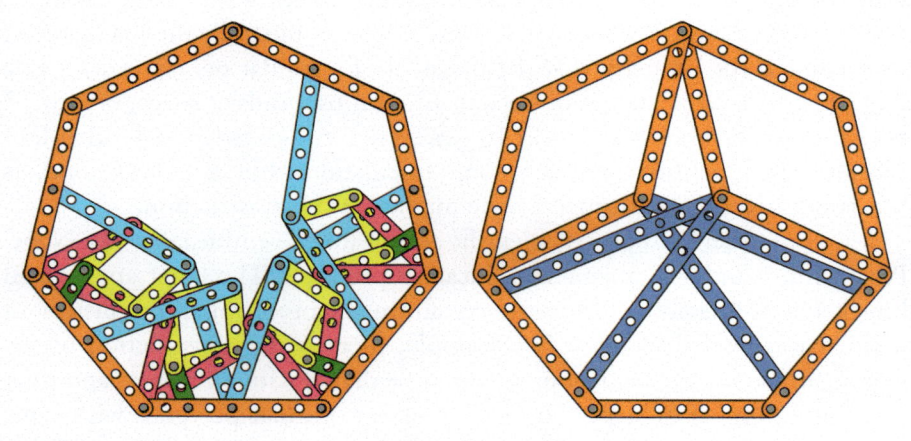

Fig. 16.4 't Hooft's constructions for the heptagon, with 43 and 15 strips

(1) How to draw a cube with double the volume of a given cube.
(2) How to divide an angle into three equal parts.
(3) How to construct a square with the same area as that of a circle.

The Greeks wanted this to be done with a compass and straightedge in a finite number of steps. Only in 1837, two thousand years later, did the French mathematician Pierre Wantzel prove that the doubling and dividing

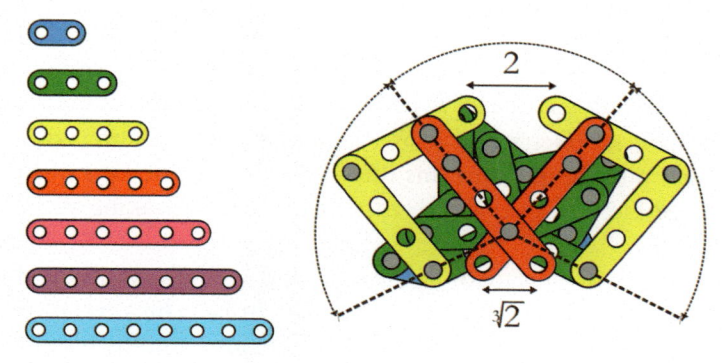

Fig. 16.5 't Hooft's construction for the cube root of 2 (with the color codes for the strips because they overlap)

problems are not possible. German Ferdinand von Lindemann gave a similar answer to the third question, in 1882. Its historical importance is illustrated by its alternative name, 'the squaring of the circle', an expression that became an idiom.

If one does not impose the conditions of only using compass and ruler in only a finite number of steps, then the problems can be solved. For example, the doubling of a cube with side 1 amounts to the construction of the cube root of 2, because a cube with side $\sqrt[3]{2}$ has a volume equal to 2. 't Hooft devised a nice construction for this, just as he did for the trisection of an angle. Later, 't Hooft magnanimously admitted that mathematician Alfred Bray Kempe (1849–1922) had already found a 'much better solution' to the latter problem in the nineteenth century—it is the great minds who admit such a thing without hesitation. It was a student of 't Hooft, Luuk Hoevenaars, who had pointed out that Meccano mathematics had been practiced for a hundred years (Fig. 16.5).

A Meccano solution for squaring the circle has not been given yet, but maybe there are interested Meccano players or LEGO®Technic Masters who can do it mechanically.

Hands-On Mathematics

There is a good didactic reason to practice mathematics with Meccano: you can do the math and look for elegant solutions to a problem without actually using formulas or doing calculations. Maybe we often practice math without realizing it. There is a comparison I often use in lectures on traditional mathematics from Africa (see my book *Africa + Mathematics*, Springer). After a lecture I was sometimes asked 'whether all that really was mathematics, without formulas or calculations?' In response I would invoke Molière's theater play *Le Bourgeois Gentilhomme* in which a 'bourgeois man' learns the difference between poetry and prose. Eventually, he exclaims with joy: 'Par ma foi! Il y a plus de quarante ans que je dis de la prose sans que j'en susse rien.' ('Incredible! For over forty years I have been speaking prose without knowing it.').

The Greeks drew their math in the sand and Euclid's 'formulas' don't even remotely look like what we are used to today. In fact, his texts often have to be 'converted' to contemporary standards. Take, for example, his definition of the 'division into extreme and mean ratio': 'A straight line segment is divided according to this ratio if the whole line segment stands to the larger part as the larger part to the smaller'. Today we would 'formulate' this as follows: call the length of the largest line segment x, and the smallest y, then that ratio is $(x + y)/x = x/y$. Since the beginning of the eighteenth century, this has been called the golden ratio.

Someone who solves sudokus is really doing mathematics, says Ingrid Daubechies, the world-famous Belgian Baroness of the Wavelets. Indeed, the puzzle is based on work by Swiss mathematician Leonhard Euler (1707–1783), but one doesn't need to know his so-called 'permutations' to have fun with a sudoku. Someone who creates geometric patterns by laying tiles practices geometry, even if he knows nothing about translations or rotations.

Likewise, when one plays with Meccano one can actually do math without even writing down a formula. It has always been my belief that someone who assembles or operates a complex machine is actually doing math. Diehard mathematicians disagree with this, sure, but as far as Meccano is concerned, I have some good support: Nobel Prize winner Gerard 't Hooft, mathematics education specialist Bart Windels and research mathematician Matthias Floré all love Meccano mathematics.

As far as I'm concerned, the definition of what is of interest to a mathematician can even be more indirect: does the work of the well-known Art Nouveau architect Victor Horta (1861–1947) involve mathematics? Architecture specialists will deny that his graceful style has anything to do with what they see as the 'rigid' science of mathematics. For Horta, they refer to plant shapes as main sources of inspiration. That is true, but those plant shapes in turn have beautiful mathematical explanations. Just think of all those graceful shapes that were drawn in descriptive geometry before the advent of CAD programs. They were realized without formulas, and yet they undoubtedly belonged to a specific field of mathematics. Professor J. Bilo (1914–2006), an influential mathematics professor at the Ghent University in Belgium, indignantly recalled the anecdote of how a worker had his application to do further studies rejected by a special examination board because 'his fingers were too rough to draw the fine lines of descriptive geometry'. Yet, perhaps that worker had already practiced more mathematics on his machines than the members of that committee on their paper forms… (Fig. 16.6).

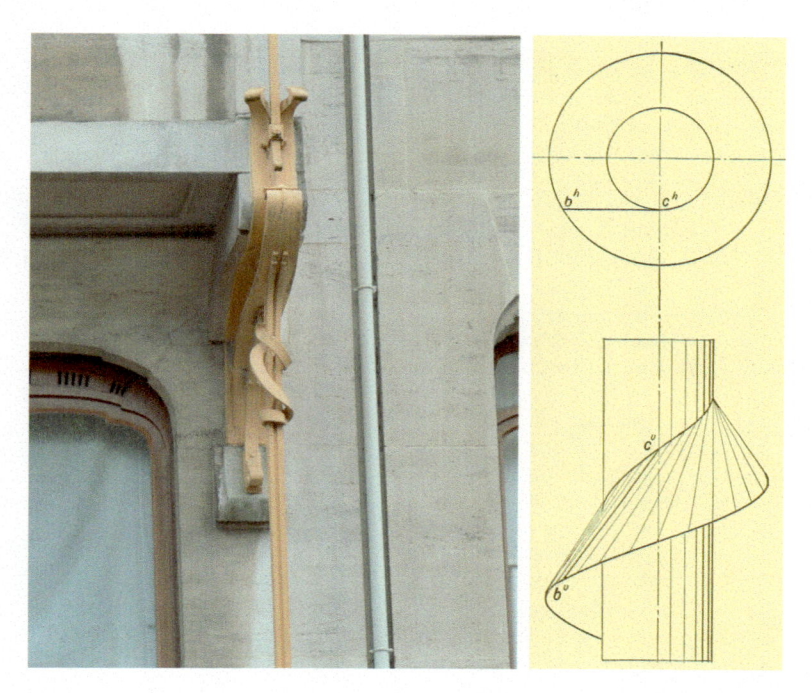

Fig. 16.6 Do you practice math when your hands design a descriptive geometry shape?

Part III

Math from Heaven

17

Mathematical Meditation

Yoga convinced many to start exercising, and perhaps Vedic mathematics can convince some to do mental arithmetic to keep a sharp mind. According to Indian teacher Tirthaji one can indeed meditate mathematically, but whether this will make computers obsolete, like his followers pretend, is another matter.

Fig. 17.1 An 'ancient Vedic multiplication', executed on a beach (*Photo* Luc Verpoort)

A Vedic Math Hype

For centuries, mathematics had a magical and even religious perception on the Indian subcontinent. Long oral traditions that can be dated back to the seventh century BC, culminated in the ideas of the sacred *Veda*, in which verses were recited on a wide variety of subjects, from large numbers to ritual geometrical constructions and astronomy. To make them easier to memorize, the rules were recited as short verse lines or sutras (Fig. 17.1).

The oldest truly mathematical manuscript was found in the village of Bakhshali near Peshawar, Pakistan. The seventy inscribed birch barks are said to date from the second half of the first millennium, though some are more precise and suggest the middle of the seventh century AD. A radiocarbon dating test in 2017 by a team from the Bodleian Library (Oxford, UK) pushed some parts of the manuscript back to the third century. It got quite some attention in the press as it featured famous mathematician Marcus du Sautoy, author of, for example, *What We Cannot Know [...]*. However, the conclusion was overruled in a most explicit way by a group of historians and linguists led by Kim Plofker:

> 'It should not be hastily assumed that the apparent implications of results from physical tests must be valid even if the conclusions they suggest appear historically absurd.'

This doesn't exclude the possibility that some of the facts explained in the Bakhshali manuscript summarize knowledge from earlier times, that is, from the first few centuries AD. The writings mark the transition to the golden age of Indian mathematics, situated between the seventh and eleventh centuries. It was a glorious era for trigonometry, spherical geometry, quadratic and cubic equations, calculations, astronomy, calculation of calendars and values of pi and Diophantine identities, and even differential calculus.

Some thirty million old manuscripts are said to have been preserved in ancient India. Not all mathematical, but all results have not yet been listed, at least from the tenth century onwards. It is interesting to take a closer look at the mathematical style they are written in, as it differs a lot from our own sometimes very boring and dry Western schoolbooks. In the twelfth century, when Europe was awakening from the Dark Ages, Bhaskara II wrote a treatise known as the Lilavati, named after his daughter. In it, he asked her questions in the following form:

> 'Fawn-eyed child Lilavati, tell me, how much is the number [resulting from] 135 multiplied by 12, if you understand multiplication by separate parts and

by separate digits. And say, beautiful one, how much is that product divided by the same multiplier?'

The Lilavati contains thirteen chapters on arithmetic and geometric sequences, plane and space geometry, combinations and equation solving. The examples are about kings and elephants and often presented as poems (Fig. 17.2):

'Whilst making love a necklace broke.
A row of pearls mislaid.
One sixth fell to the floor.
One fifth upon the bed.
The young woman saved one third of them.
One tenth were caught by her lover.
If six pearls remained upon the string
How many pearls were there altogether?'

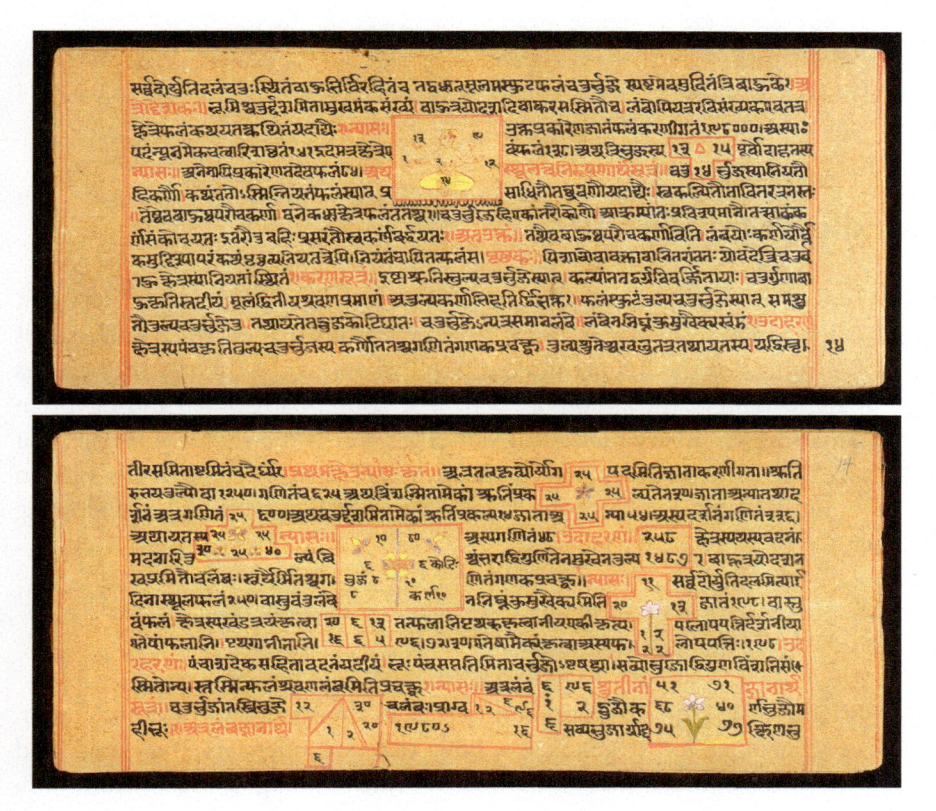

Fig. 17.2 Two pages from the Lilavati

The conclusion of Bhaskara's mathematical book is romantic too:

> 'Joy and happiness are indeed ever increasing in this world for those who have Lilavati clasped to their throats, decorated as the members are with neat reduction of fractions, multiplication, and involution, pure and perfect as are the solutions, and tasteful as is the speech which is exemplified.'

No doubt, Indian mathematics is interesting and of a high level and pleasing to read. It would be enjoyable if some of the results were used in the original versions in educational texts, in which the Eurocentric world view still predominates. Here is but one example, beautiful because of its simplicity and because of the unexpected answer (and no, no data are missing):

> There are two columns. A thread connects the top of one column to the base of the other. Another thread connects the top of the other column to the base of the first. From the point where both ropes intersect, another rope is tied, hanging down and touching the ground. Now then, know that the length of this rope is half the harmonic mean of the heights of the columns.

That is, if h_1 and h_2 are the heights of the columns, then the length h of the rope is $1/(1/h_1 + 1/h_2)$, or else: $\frac{1}{h} = \frac{1}{h_1} + \frac{1}{h_2}$.

Magical Mathematics

The Indian subcontinent can be proud of its heritage in science, philosophy, and culture. There is, for instance, the Pythagorean theorem, which states that the square of the hypotenuse equals the sum of the squares of the sides of the right triangle, the famous $a^2 = b^2 + c^2$. Sure, Babylonian and Egyptian mathematics gave many numerical values for which the theorem is true, but the first true mnemonic is due to Katyayana (200–140 BC). In his version of the Sulba sutras, which go back to about 800 BC, there is the following sentence:

> 'The diagonal rope of an oblong (rectangle) produces both which the flank and the horizontal produce separately.'

Note the word 'squared' does not appear in the statement of the theorem, but of course, a lot depends on how that statement is translated and interpreted. We must keep in mind we are reading a translation of a thousand-year-old language from an entirely difference culture.

And yet, too great an enthusiasm for a multicultural approach to mathematics can lead to mythical exaggerations too. For instance, a lot of good will is needed when reading the so-called 'ancient Vedic mathematics' explained in a book by a certain Jagadguru Swami Shri Bharati Krishna Tirthaji Maharaja. It was published five years after Tirthaji's death, in 1965, partly because there was no certainty that the actual content reflected the Vedic scriptures. But thanks to some effective propaganda, assisted by some nice viral YouTube movies, the impression was created that ancient Indian mathematics knew a 'magical multiplication method'.

More Sutras

The Vedic scriptures contain sixteen rules or sutras and thirteen similar sub-sutras of a mathematical nature, according to Tirthaji. He made it his life's work to show that these sutras are the keys to the secret ancient mathematical knowledge hidden in the Indian Vedas. The mathematical aphorisms would have been contained in the part called *Parishishta*, an appendix of the *Atharva Veda*, but even that is doubtful, because no one, including Tirthaji, could ever show exactly where they were mentioned. The Sanskrit in which the texts are written does not make it easy to confirm or deny it, but even if the passages in question were there, it would remain uncertain that mathematical calculations could be derived from them in the way Tirthaji proposed.

For example, the second sutra reads:

All of 9 and the last of 10.

The proofreader of this book, Christopher Dunkley, pointed out this may refer to a way of performing subtractions. Indeed, to calculate, for instance, 100,000—98,765, one simply subtracts all the digits in the second number individually from 9 ("all of 9"):

$$9 - 9 = 0, 9 - 8 = 1, 9 - 7 = 2, 9 - 6 = 3$$

and the last number from 10 ("the last of 10"):

$$10 - 5 = 5.$$

This gives the answer: 1235. Of course, this only works if the number of zeros in the first number is equal to the number of digits in the second.

This simple principle could also refer to ways of multiplying numbers. Take 88 × 96 as a straightforward example. Consider first that

$$88 = 100 - 12 \text{ and } 96 = 100 - 4.$$

Then:

$$88 \times 96 = (100 - 12) \times (100 - 4)$$
$$= 10,000 - 1200 - 400 + 48 = 8448.$$

Via the "all of 9 and the last of 10" principle, tabulate this process.

Number	Difference from 100
88	−12
96	−4
Sum of both, above 100: 84	Product: 48
Thus: 88 × 96 = 8448	

Or else, for 1038 × 1006:

Number	Difference from 100
1038	38
1006	6
Sum of both, above 100: 1044	Product: 228
Thus: 1038 × 1006 = 1,044,228	

In modern notations, this multiplication corresponds to:

$$1038 \times 1006 = (1000 + 38) \times (1000 + 6)$$
$$= 1000 \times (1000 + 38 + 6) + 38 \times 6$$
$$= 1000 \times (1044) + 228 = 1,044,000 + 228 = 1,044,228.$$

That is no 'higher' mathematics, but a simple calculation any high school student can perform using the formula $(x - b) \times (x - c) = x \times (x - c - b) + b \times c$, where $x = 100$.

Another sutra reads:

Vertical and crossed.

This could refer to the direction in which the components of a multiplication are worked out if the numbers to be multiplied are placed in columns. So, applied to 36 × 53 = 1,098, for example, the units are calculated vertically (6 × 3) and the tens in a crossed formation, so (3 × 3) and (6 × 5), and the finally the hundreds are calculated vertically again (3 × 5). This translates into the following drawing and table (Fig. 17.3):

Tens			Units	
3			6	18
	×			+ 39
5			3	+ 15.
Product: 15	3 × 3 + 5 × 6 = 39		Product: 18	= 1908

In this example, the 15 brown shells, the 9 + 30 = 39 black and the 18 white shells lead to 18 + 390 + 1500 = 1908, and that is indeed 36 × 53.

In our modern notations, this corresponds to

$$36 \times 53 = (3 \times 10 + 6) \times (5 \times 10 + 3)$$
$$= 3 \times 5 \times 100 + 10 \times 3 \times (3 + 6 \times 5) + 6 \times 3$$
$$= 1500 + 390 + 18 = 1908.$$

With larger numbers the method is more complicated but even more magical, though it always comes down to the same principle. Indeed, any high school student can check how, for instance, 2,341 × 683 = 1,598,903

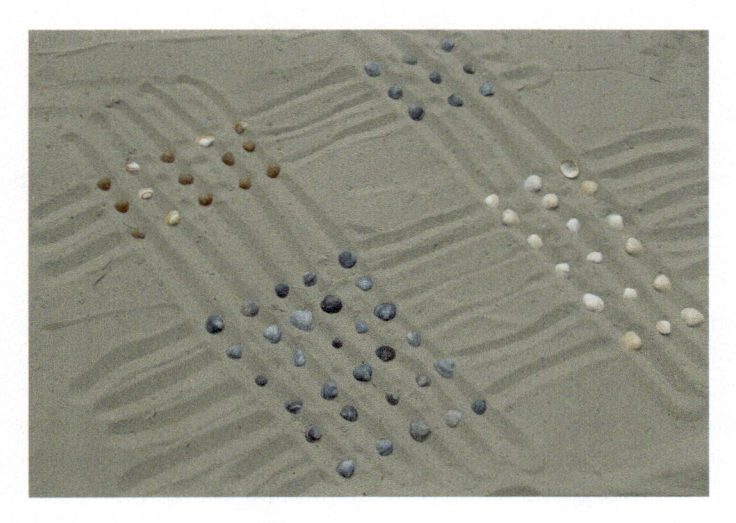

Fig. 17.3 Vedic multiplication of 36 × 53

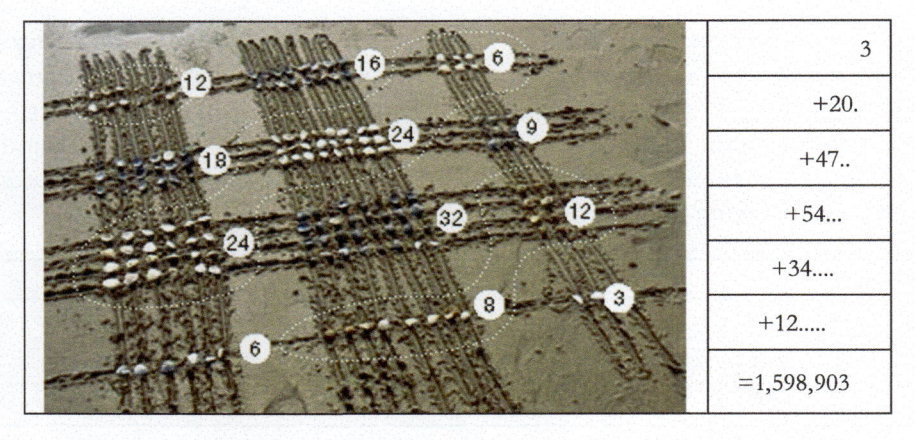

Fig. 17.4 Vedic multiplication of 2341 × 683

is performed by evaluating the product $(ax^3 + bx^2 + cx + d) \times (gx^2 + hx + i)$, where $x = 10$ (Figs. 17.1 and 17.4).

In this example $12, 16 + 18 = 34, 6 + 24 + 24 = 54, 9 + 32 + 6 = 47, 12 + 8 = 20$ and 3 are the key numbers:

$$2341 \times 683 = \left(2 \times 10^3 + 3 \times 10^2 + 4 \times 10 + 1\right) \times \left(6 \times 10^2 + 8 \times 10 + 3\right)$$
$$= 2 \times 6 \times 10^5 + (2 \times 8 + 3 \times 6) \times 10^4 + (2 \times 3 + 3 \times 8 + 4 \times 6) \times 10^3$$
$$+ (3 \times 3 + 4 \times 8 + 1 \times 6) \times 10^2 + (4 \times 3 + 1 \times 8) \times 10 + 1 \times 3$$
$$= 12 \times 100,000 + 34 \times 10,000 + 54 \times 1000 + 47 \times 100 + 20 \times 10 + 3$$
$$= 1,598,903.$$

The calculations are nice and correct. With some practice, they can be performed easily; there are no features of higher mathematics as they do not go further than advanced mental arithmetic. Also, the question remains whether the formulas really follow from the sutras 'all of 9 and the last of 10' or 'vertical and crossed'. This connection transcends most mathematical karmas.

Modern Fantasies

It is more likely that the alleged ancient Vedic mathematics are the modern inventions of Tirthaji. As a kind of mathematical guru, he pretended to have derived 'all mathematics for scientific calculation' from some mysterious verses, just as other gurus derive 'all wisdom of life for the human karma' from other verses. Tirthaji himself was a teacher at a traditional Indian school called the Govardhana Matha. He was highly regarded because his family came from a higher class. He had a good education, which was not a given in early twentieth century India. In 1903, he would set a record by obtaining the highest score for Sanskrit, philosophy, English, mathematics, history, and science in the examination commission of the American College of Sciences in Bombay!

Thus, he not only excelled in mathematics, but also in the study of Sanskrit, and that perhaps led him to the interpretation of the ancient Indian scriptures. After his studies, between 1908 and 1925, he would spend several years meditating about the Veda philosophy, sometimes by retreating to the jungle, occasionally interrupted by 'worldly' assignments as a teacher or counselor. On 4 July 1919, he saw the light and began to preach the Vedic philosophy as a swami, an honorific title he received for having chosen 'the path of renunciation'. He would start spreading his message of peace, harmony, brotherhood and… mathematics.

Tirthaji gained more and more supporters, including a secretary of state in finance and some university graduates. In 1958, he went to the United States and Great Britain as the first traditional Indian teacher to be awarded the title of Shankaracharya. After his death, in 1965, a chair of Vedic Mathematics was established at Banares Hindu University, by a wealthy mathematics enthusiast.

His efforts were not in vain. Some of his disciples expressed their admiration as follows:

'Since the advent of the calculator, we occupy our brains less and less with calculations. The practitioners of Vedic mathematics believe that we are depriving our brains by doing so. They regard mental arithmetic as healthy brain gymnastics, which also offers a lot of fun. Going back to Indian traditions, they developed a system of calculation methods with which they often find the outcome surprisingly quickly, on their own'.

Other Vedic adepts go even further and state that their 11,000-year-old mathematics (which, in fact, is at most 70 years old), would even outdate computers:

'When everyone becomes an expert in Vedic mathematics, computers and calculators will become obsolete and the entire computer industry will go bankrupt'.

Tirthaji's 'wisdoms', however, have an authenticity problem, as confirmed by many Indian scholars. Georges G. Joseph was quite moderate in his acclaimed *The Crest of the Peacock*, but his book dates from 1992, when the Vedic tale had not yet received any media attention. Shrikrishna G. Dani, however, of the Tata Institute of Fundamental Research, says that Tirthaji's pseudoscientific teaching poses a danger for the image of Indian science. In *Myths and Reality: On 'Vedic Mathematics'* he wrote that the Tirthaji's teaching is in no way based on old Indian mathematics, and that it also contains but a few mental arithmetic techniques, be it embellished by mystical enchantments.

For the most part, it repeats the arithmetic from the old textbooks Tirthaji used himself at school. Maybe he made up some tricks himself, or maybe he learned some from older mathematicians. But contrary to what 'the teaching' claims, 'studying for 8 months, 2–3 h a day, on an average, instead of 15–20 years under the existing systems at Indian and other universities is *not* sufficient 'to acquire the entire mathematical knowledge'. More accurately stated, 8 h is enough time to review the whole of Tirthaji's mental arithmetic lessons.

The authenticity problem of the Vedic method of multiplying numbers by intersecting lines is perhaps best illustrated by the mixed messaging it has been subjected to on social media (the medium by which it got its popularity): the same method also circulates as a 'Japanese method'. Its roots are so vague the author was not surprised to hear it during a conference on 'African mathematics' as well. It indeed looks quite spectacular when presented swiftly on a blackboard.

In addition, there are many other systems for fast mental calculations. Ukrainian engineer Jakow Trachtenberg developed one to occupy his mind while in a Nazi concentration camp. In modern times too, mental arithmetic may be helpful to keep the mind sharp.

On the positive side, it is true that Vedic mathematics draws attention to mental arithmetic, even though, historically, its authenticity is not proven, and its attractive method exaggerates its mathematical complexity. Its unusual and fun style could convince many to turn to mathematics, just as yoga persuaded many to do physical exercise. Some apps or software programs propose mental fitness exercises, but for Tirthaji's arithmetic contemplation it suffices to draw some lines in the sand and count shells placed at their intersections. Just meditate mathematically, humming humm times hummm is hummmmmm…

18

Did Newton's Apple Fall First in India?

In 2018, G. G. Joseph sent out a warning: Hindu nationalists abused his work to state that Isaac Newton had copied his gravitational theory from India. This exaggeration was a pity, for some spectacular formulas from India did indeed precede Newton and perhaps reached him through Jesuit missionaries.

Fig. 18.1 Statue of Isaac Newton at the South-India Regional Science Centre in Calicut

Did Jesuits Import Calculus from India?

George Gheverghese Joseph, professor at the University of Manchester, became famous in the field of the history of mathematics because of his book *The Crest of the Peacock: Non-European Roots of Mathematics*. He discussed non-western mathematics from North and South America, from ancient Egypt, Mesopotamia, China and India, as well as Islamic contributions, all from periods that predate contact with the West. The book was a reaction to the usual historiography that lets mathematics begin with the ancient Greeks from about 600 BC to 400 AD, being reborn in the Renaissance of the sixteenth century, sometimes adding that the Islamic world was responsible for partially transferring Greek knowledge to Western Europe (Fig. 18.1).

Joseph stressed the importance of the Congolese Ishango rods (the first tally sticks), Yoruba shell counting (multiplications in base 20), Egyptian papyri (problems of a practical and recreational nature), clay tablets from Mesopotamia (numerical examples of the Pythagorean theorem), Indian mathematics (from the invention of the zero to more advanced mathematics) and China (square and cube roots, Pascal's triangle). He proposed an alternate route showing how mathematics reached Europe from different parts of the world, with an important Persian contribution as well. Note Joseph also paid attention to the mathematics of the Maya and the mathematics originating from Southeast Asia (Fig. 18.2).

The course of Joseph's own life probably explains his commitment to these 'multicultural mathematics', a topic that barely existed when he wrote his book. He was born in Kerala, the area on the western side of the southern tip of India, and moved to Mombasa, Kenya, at the age of nine. He pursued his higher education at the universities of Leicester and Manchester, doing research in applied mathematics and statistics. Later, he focused on cultural and historical aspects of mathematics, with particular emphasis on non-European dimensions and their relevance for mathematics education (Fig. 18.3).

From time to time, Joseph returns to his native region and although he was initially very pleased with the reception and the media attention caused by this visit, he was startled by certain interpretations of his work. Certain Hindu nationalist blogs exaggerated some of his statements adding all kinds of claims. Some of these blogs got tens of thousands of hits in no time (see for instance the blog with the telling title: 'OMG!!! Manchester University confirms that Isaac Newton stole gravity theory from Hindus').

In March 2018 Joseph felt compelled to send a message to his circle of historians of mathematics and friends. He concluded: 'I strongly object to

18 Did Newton's Apple Fall First in India? 169

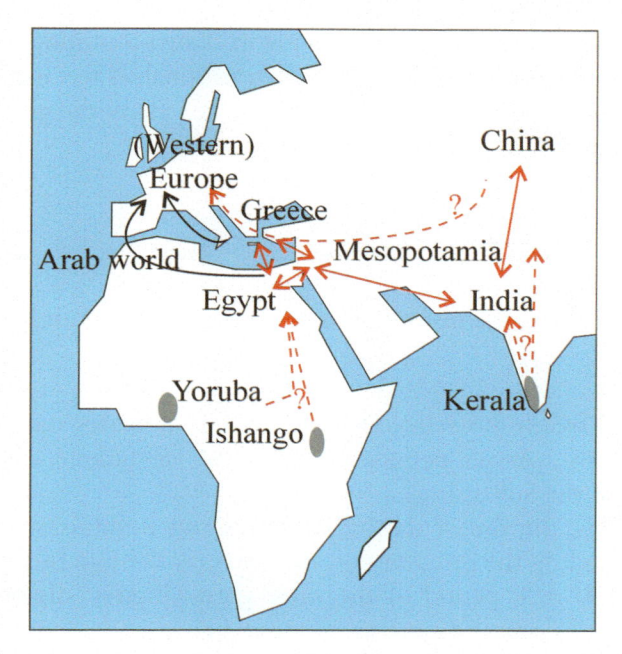

Fig. 18.2 The classic road for the history of mathematics followed the black arrows; G. G. Joseph added the red ones

Fig. 18.3 George Gheverghese Joseph

my writings being misconstrued by people who are promoting views which I find as anathema.' But why did these Indian nationalists feel inclined to wave the mathematical flag accusing Isaac Newton, for many the greatest scientist ever, of being a thief?

Mathematics from Kerala

According to Joseph, the history of mathematics in India can be divided into six periods:

- 3000–1500 BC, Indus valley: development of measures and weights.
- 1500–800 BC, Aryan period: astronomical, arithmetic, and geometric questions in the Sulbasutras.
- 800–200 BC, the rise of the Indian states and Buddhism: Vedic mathematics, including the discovery of, for instance, the binomial theorem.
- 200 BC–400 AD, period of the three great empires: discovery of zero, quadratic equations and systems, negative sign.
- 400–1200 AD, Gupta dynasty: 'classical period', with systems of quadratic equations, triangular geometry, tables for the sine.
- 1200–1600 AD: decline of mathematics in the North under Muslim dynasties, though still flourishing in the isolated Southern Kerala.

The dating of the Bakhshali manuscript was discussed in the previous chapter: though the birch barks themselves seem to have been written in the seventh century, some parts refer to results known between 200 BC and 400 AD. As explained, this dating got quite some media attention, partially because the discovery of zero was the main topic. Yet, the mathematics of the Kerala School is at least as spectacular, though it remains rather unknown, perhaps because its mathematics is less accessible. For instance, a formula 'officially' discovered by the Scot James Gregory (1638–75) in 1667, was already known to Madhava of Sangamagramma (1340–1425), that is, about three centuries earlier.

Today, Gregory's formula is a standard calculus topic. Consider a right triangle OAB with a hypotenuse 1, forming an angle t in O. The latter is expressed in units called 'radians', where 180° = pi radians and thus 1 rad = 180°/pi. The sine of t is the length of the opposite side AB, and the cosine of t is the length of the adjacent side OA. The tangent of t is the quotient of both, and thus the length of opposite side of the angle divided by the length of the adjacent side. Now Gregory's formula states that:

$$t = tan(t) - \frac{tan^3(t)}{3} + \frac{tan^5(t)}{5} - \ldots$$

Madhava of Sangamagramma discovered this formula much earlier, and so his name should be linked to it too (Fig. 18.4).

Similarly, the formula

$$\frac{\pi}{4} = 1 - \frac{1}{3} + \frac{1}{5} - \ldots$$

should be called 'formula of Madhava—Nilakantha—Leibniz', though it is always attributed only to the Sorbian-German Gottfried Leibniz (1646–1716). It would be fair to include a third name, that of another Indian mathematician, Nilakantha Somayayi (1445–1545), though he attributed this result to Madhava. Note the formula follows from the previous one by setting $t = pi/4$.

In 1676, Isaac Newton sent two formulas in a letter to the Royal Society, one for the sine of an angle and another for the cosine:

$$sin(t) = t - \frac{t^3}{3!} + \frac{t^5}{5!} \ldots \text{ and } cos(t) = 1 - \frac{t^2}{2!} + \frac{t^4}{4!} \ldots$$

They are among the most beautiful formulas in mathematics, but in fact they were discovered three hundred years before Newton, by the aforementioned Madhava and thus they should be called 'formulas of Madhava—Newton'.

The complexity of the formulas for non-mathematical readers may explain why this spectacular fact remains virtually unknown, in contrast to the press attention for a known fact about the number zero. Or perhaps the complex

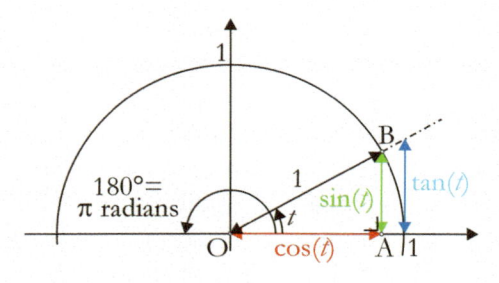

Fig. 18.4 The sine, cosine and tangent of an angle t

formulas seem less reconcilable with the common perception many have about India from 500 years ago, as they seem to come straight out of a modern engineering course. The association of Indian mathematics with the Hindu-Arabic numerals, with the number zero, or with nice and fun drawings seems more acceptable. In addition, it is not easy to represent this mathematics correctly because of the many linguistic problems involved. For instance, Madhava calculated that the circumference of a circle with a diameter of 900,000,000,000 is approximately equal to 2,827,433,388,233, but he expressed this number as follows: Gods (33), eyes (2), elephants (8), snakes (8), fires (3), three (3), qualities (3), vedas (4), naksatras (27), elephants (8), and arms (2).

Not only did the Kerala mathematicians discover the given formulas several hundreds of years before their Scottish and English counterparts, but one could also even argue the Kerala school founded the field of mathematical analysis, as there are other results supporting this claim. Moreover, there are indications that Jesuits might have transferred certain writings from Kerala to Western Europe. In 1540, the Spanish Jesuit missionary Francis Xavier (1506–52) arrived in Goa, a former Portuguese colony in India. After 1578, other Jesuits with a special interest in mathematics embarked there, such as Matteo Ricci (1552–1610), who trained in mathematics at the Collegio Romano of Christoph Clavius (1538–1612).

Other 'mathematical missionaries' were Johannes Schreck (1576–1630), who had worked with the French mathematician François Viète (1540–1603), and Antonio Rubino (1578–1643). They would go to Kerala, though it is about 750 km away, south of Goa. It is therefore not impossible that the formulas for the sine, cosine, and tangent and for pi/4 came from India—though no direct evidence of this conjectured transmission of knowledge is yet available.

What Went Wrong?

However, it is another matter to conclude from this that Newton's entire theory of gravity was 'stolen' from India. Indeed, when references to certain sutras are considered to support this, matters become vague. As explained in the previous chapter, a sutra is an aphorism, a short saying or verse. It can be about the most diverse subjects. The texts *Ideas of the Holy Veda* (7th–fourth century BC) summarize the long oral traditions from India of those times. In the *Aryabhatiya* (499 AD) there are 33 sutras with mathematical content, and Bhaskara I (600 AD) illustrated a few of them with numerical examples.

Some contain names for very large numbers, for geometric constructions, or astronomical considerations.

The previous chapter described how some of these sutras were interpreted as mental computational techniques 'that would allow us to ban computers'. Quite an overstatement, but this is exactly what is now happening with Newton's gravitational theory. This time pseudo-scientists turn to the Vaisesika sutras, written between the sixth and second century BC. The author is a certain Kanada, but he is also called Kashyapa, sometimes with the title of 'rishi', which stands for 'inspired poet'. His sutras contain theories about the creation of the universe, in which he explained it is made of atoms. The verses are indeed reflections about physics—so far, so good. However, deducing the laws of Newton from them is another matter. Moreover, as the Vaisesika sutras were written between the sixth and second century BC and the discussed formulas of Keralan mathematics in the fifteenth and sixteenth century, there are at least 17 centuries between those sutras and those formulas. It is therefore unlikely that those general sayings were combined with the purely mathematical formulas, as Newton did later on. Of course, that Newton's apple first fell in India produces a juicy story, into which Indian nationalists and lovers of exotic pseudoscience can eagerly bite (Fig. 18.1).

19

Allah's Nonagons

A nonagon is a polygon with nine vertices on a circle and nine identical edges, but it cannot be drawn with a compass and ruler in a finite number of steps. Yet the Istanbul Hagia Sophia is decorated with nonagonal patterns, just as some other mosques are. They puzzle mathematicians and Muslims.

Fig. 19.1 The Hagia Sophia in Istanbul with intertwined pattern of hexa- and nonagons

Mathematical Embellishments

Islamic art provides beautiful examples of regular polygons, formed by fitting tiles together. Such regular polygons are geometric figures with vertices on a circle and edges of the same length, mutually forming the same angle, two by two. An equilateral triangle, a square and a hexagon are the classic examples with which many pupils try out their ruler and compass. What can be more fun than creating beautiful, regular figures by simply drawing circles and straight lines (Fig. 19.1)?

It is also possible to make them without ruler and compass, using tiles and fitting elementary pieces together as in a puzzle. The practice of incorporating such abstract patterns on buildings originated in the Muslim world during the Caliphate of the Abbasids (750–1258), the Islamic empire that included, at its height, North Africa, the Arabian Peninsula, Southern Europe, the former southern Soviet states and present-day Iran, Afghanistan, and Pakistan. The graceful patterns with plants and flowers that characterized architecture under the Umayyads (660–750) had to make way for geometric shapes. Since then, the myth has been perpetuated that non-abstract figurative patterns are sacrilegious, though this is not mentioned in the Quran (or Koran).

From the eighth century onwards, six- and eight-pointed stars began to appear in all kinds of patterns, as well as an occasional twelve-pointed star. From the eleventh century onwards, more stars appear with five, seven, nine, ten, eleven, twelve or thirteen vertices. In Persian Safavid architecture (1501–1763) decorating the eight- and ten-pointed stars with plant patterns, and sometimes with calligraphic inscriptions, was again allowed. Between 1526 and 1858 the Islamic Mughal dynasty in India also introduced the practice of using geometric patterns as decorations (Fig. 19.2).

Surprise: Nonagons

Not only Muslims but also mathematicians are thrilled with the geometric preference for nonagons, and the construction of these figures has been studied in detail, repeatedly. Yet some years ago Spanish mathematician Antonia Redondo (IES Bachiller Sabuco, Albacete) could still surprise experts attending a conference at the Middle East Technical University (METU) in Ankara, Turkey, when she found nonagons woven by hexagonal patterns. She gave examples from the Hagia Sophia in Istanbul and from the Selimiye Mosque in Edirne (Turkey). This is curious indeed, because drawing polygons by compass and ruler was a mathematical subject by predilection since

Fig. 19.2 Scenes from nature are featured on the great Mosque (705) in Damascus

the Greeks, and Islamic scholars had further contributed to this topic. What then could still be so remarkably amazing about intertwining hexagons producing nonagons? Everyone 'sees' them, of course, but what could a closer mathematical look reveal? (Fig. 19.3).

Drawing polygons, especially with a compass and a ruler in a finite number of steps, is a long-standing preoccupation in mathematics, ever since the Greeks started trying to do it. Drawing an equilateral triangle or a square is not so difficult, but a regular pentagon is already a challenge. A regular hexagon is again simple, but a heptagon is impossible. And in this way it continues, in a seemingly haphazard combination of possible and impossible: an octagon is constructible, a nonagon and hendecagon are not, a decagon and dodecagon are, but not so regular polygons with thirteen and fourteen vertices, while those with fifteen and sixteen vertices can again be drawn. The case of the heptadecagon (17 sides), which can be constructed, brought eternal fame to Carl Friedrich Gauss, in 1796. Later he even proved, more generally, that only regular polygons with n vertices can be drawn if n is the product of a power of 2 and distinct Fermat prime numbers. Five of these Fermat numbers are known: 3, 5, 17, 257 and 65,537 (Fig. 19.4).

Fig. 19.3 The Selimiye Mosque in Edirne has non-geometric figurative images and a minbar with intertwined hexa- and nonagons

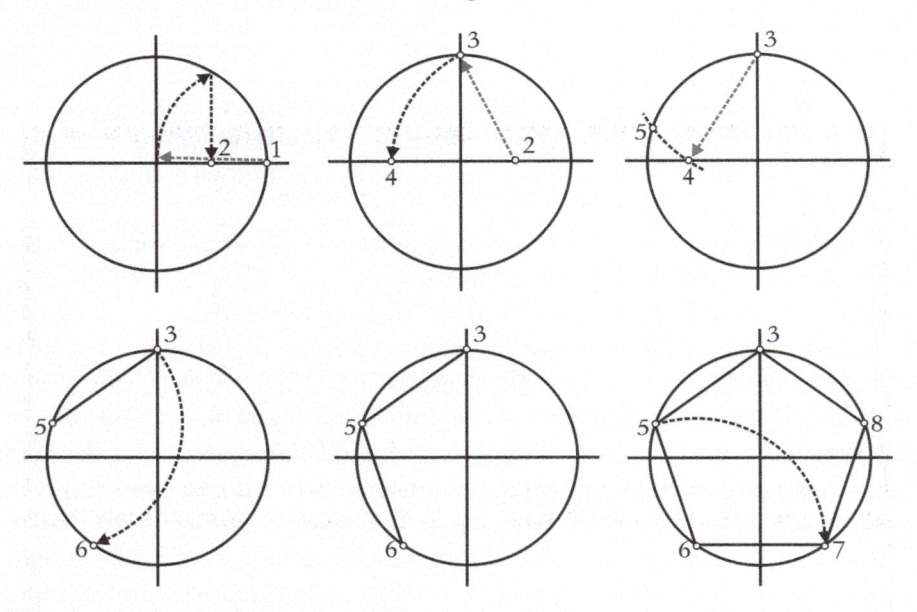

Fig. 19.4 A regular pentagon can be constructed by a ruler and a compass in a finite number of steps

Thus, it is impossible to draw a heptagon or a nonagon with a compass and a ruler in a finite number of steps, but that doesn't mean good approximations can't be made. In the late eleventh century, Muslim architects discovered constructions for almost regular hepta- and nonagons in which the error is indistinguishable by the naked eye. Literature mentions these nonagons, albeit rarely. Nonagons woven by hexagons seem mainly the work of Mimar Sinan (1489–1588). His disciples spread their knowledge, perhaps as far as India, where the pattern was later found as well.

First, Redondo investigated the possibility of drawing this hexa- and nonagonal pattern via proportions. This is a usual way, for modern mathematicians, as the drawing of regular polygons can be based on the simple relationships between edges and diagonals. The golden ratio is well known: it is the number $\phi = 1.618...$, the length of a diagonal in a pentagon with side 1. It is one of the solutions of the equation $\phi = (1 + \phi)/\phi$ or $\phi^2 - \phi - 1 = 0$ and has remarkable properties. For instance, the quotients of the consecutive numbers of the Fibonacci series 1, 1, 2, 3, 5, 8, ..., where each term is the sum of the two previous ones, tend to the golden ratio.

The study of heptagons and nonagons shows that one can also discover interesting relationships for the lengths of diagonals in a heptagon and a nonagon. In a heptagon with side 1, two diagonals can be distinguished with identical lengths, $\rho = 1.80194...$ and $\sigma = 2.24698...$ and they obey the equalities $\rho = (1 + \sigma)/\rho$ and $\sigma = (1 + \sigma)/\rho$. In a nonagon there are three, $\alpha = 1.87938...$, $\beta = 2.53208...$ and $\gamma = 2.87938...$ (so indeed, $\gamma = \alpha + 1$). Now, $\alpha = (1 + \beta)/\alpha$, $\beta = (\alpha + \gamma)/\alpha$ and $\gamma = (\beta + \gamma)/\alpha$. The numbers ρ, σ and α, β and γ can also be linked to generalized Fibonacci series. Of course, knowing the lengths of those diagonals allows the construction of those polygons (Fig. 19.5).

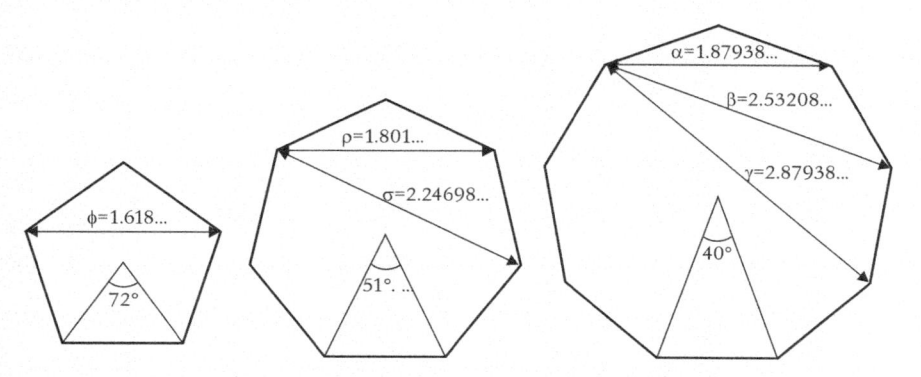

Fig. 19.5 The golden section in a pentagon (left) and the equivalents for hepta- (middle) and nonagons (right) with edges of length 1

However, when laying tiles, these decimals, expressions and equations seem rather elaborate. And thus, Redondo also explored a method based on the trisection of the 60 degrees angle. In this way, an angle of 40 degrees can be drawn from the center to one side of a hexagon, and 360°/9 = 40°. There is even a theoretical motivation to investigate this relationship, considering that the Persian scholar Abu Rayhan al-Biruni (973–c.1050) showed how to find the '40-degree chord' by solving the equations $x^3 - 3x - 1 = 0$ and $x^3 - 3x + 1 = 0$.

Yet again, it seems improbable a tile layer would solve third order equations to make tile arrangements on a floor or wall. That is why Redondo finally suggested two practical options. Perhaps the tilers had found out, by experience, that they could easily get an angle of 38.2 degrees in a hexagonal honeycomb pattern by using intersecting lines connecting certain points in the pattern. This angle is close enough to 40 degrees for practical purposes, so that a nonagon can be constructed, and thus the whole pattern (Fig. 19.6).

Another possibility was that the tile layers started with some basic puzzle pieces Redondo called 'star', 'bow' and 'house'. Putting puzzle pieces together in a certain way produces the desired pattern. Again, the nonagons are only approximately regular, but the error is not noticeable when the geometry is done with tiles and plaster (Fig. 19.7).

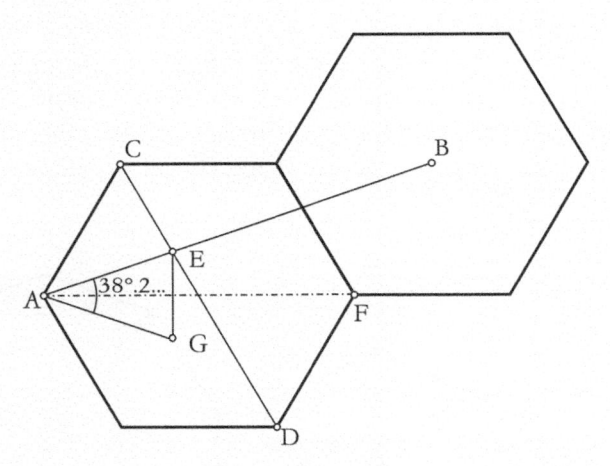

Fig. 19.6 The lines AB and CD intersect in E and together with G (symmetric to E with respect to AF), EAG forms an angle of 38°0.2…, a good approximation for the 40° angle allowing the construction of a nonagon

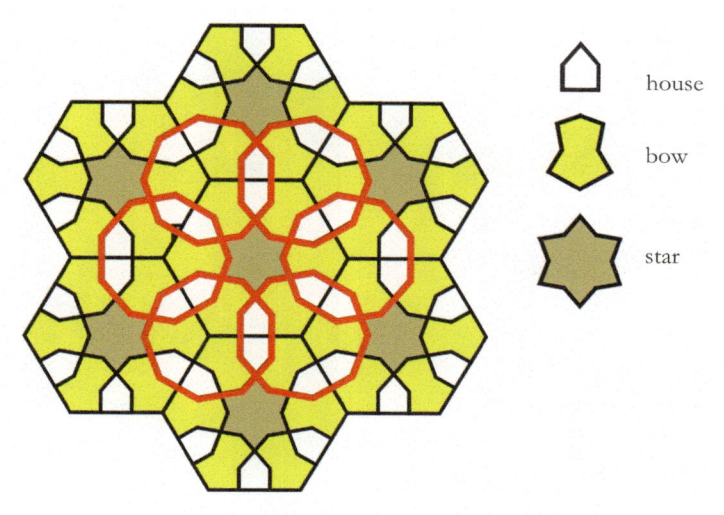

Fig. 19.7 A puzzle system to weave nonagons using hexagons

Mystery

The hexagonal woven nonagons raise several questions. Redondo had come to her assumptions through mathematical arguments and intuitively defended the puzzle hypothesis. In the METU conference, she was supported by historians who asserted that the so-called Topkapi scrolls (Istanbul, fifteenth century) contain 114 examples of similar drawing schemes, for example for triangular, quadrangular, and pentagonal patterns. Peter Lu (Harvard University) also defended the 'gereh tile' method (the word 'gereh' means 'tile' in Farsi, so the name 'gereh tile' is a tautology) in which patterns are formed puzzle-wise, following certain given lines. The system would have had five standard tile puzzle pieces and have been in use since the 1200 s. However, it remains unlikely they would help in the creation of the heptagonal patterns and hexagonal woven nonagons.

Another enigma is the preference for nonagons. After all, nine has no special meaning in Islam. There are 99 pearls on a traditional prayer wreath, called 'sibha', 'subha', 'misbaha' or 'mashaba' (in Arabic), 'tasbih' (in Persian), 'tespih' (in Turkish and Bosnian). This number is seen as 3 times 33 and stands for 3 sequences of glorifications or names for Allah. And yet the hexagonal woven nonagons in the Selimiye Mosque in Edirne stand on a minbar or 'sacred staircase', a special part of the interior. These staircases are never used as such, to go to another floor, but stand as a reference to the heavenly world above. Only on special occasions will a dignitary go up to the nonagonal patterns (Fig. 19.8).

Fig. 19.8 Mishaba with 99 beads (left) and with 33 beads (right)

Perhaps the makers of the exceptional nonagons just wanted to show how good they were and how well thought out their ornaments were. If the neighboring city has a heptagonal pattern, the next mosque will have a nonagonal one. Christians glorified their faith by building the most pointed Gothic church. Muslims looked for the most surprising polygon pattern, a choice that is rewarding from many mathematical angles.

20

Is Mathematics Halal?

For about 400 years, the exact sciences are 'permissible' to some fundamentalists in Africa, the Middle East, and Southeast Asia. What do they accuse mathematics of, 'for heaven's sake'?

Fig. 20.1 Madrassas or Quranic schools read the Quran all day long—little time is spent on math or science

Mathematics is Evil

'Boko Haram', literally means: 'Western education is forbidden'. It is the name for a Nigerian rebel group fighting against mathematics education. The members fear that children will lose their Islamic roots due to the influence of modern schools. Their leader Mohammed Yusuf (1970–2009) defended a religious society without Western influences: 'Democracy and the current education system must change. Otherwise, the war, which is about to begin, will last a long time'. It turned out he was right about the latter part.

The main issue of 'Boko Haram' is a ban on modern schools, which are increasingly replacing Islamic schools today. The traditional teachings under a tree or in a house explained to students how to write and recite the Quran and, at a later age, gave them insights into Arabic literature and theology. However, English, mathematics and sciences in general, are not on the program. Some resistance to biology classes because of explicit sexual facts might sound recognizable, as might the rejection of history classes advocating the theory of evolution, or learning English, the language 'of the American enemy'. But what 'the hell' could cause the opposition to mathematics? Does not mathematics use abstract patterns like those decorating mosques? Doesn't the term 'Hindu-Arabic numbers' indicate a clear link between mathematics and the Arab world? By the way, the word 'cipher' refers to the Arabic 'sifr' which means zero. And when doing a long multiplication, the numbers are still written one below the other from right to left, as in Arabic writing.

And yet, many Muslim schools do not teach the exact sciences. In most Asian madrassas or Quranic schools, not one cipher is written unless it is a number from a verse of the Holy Book. Not surprising Arab universities do not count many first-class mathematicians: for 400 years there has not been a single Arab top mathematician—this in great contrast to, for example, the innumerable Jewish names adorning the mathematical list of honor (Fig. 20.1).

Bad Grades

Since 2003, the Singapore government has been exerting pressure on the six madrassa schools in the country, because the government concluded that their teaching in non-religious subjects is poor. The government required all madrassas to attain a given minimum level, otherwise they would be denied the right to organize their curriculum. Some madrassa schools received special support from the government and were equipped with the most modern technical devices. Their educational model, in which the students start the day with an extensive prayer session, and then study science and mathematics, is reluctantly implemented in Indonesia and the Philippines as well. It contrasts with most of the traditional madrassas in Southeast Asia, where students do nothing but memorize the Quran, all day long. Usually these holy texts are written in a language that is incomprehensible to them, that is, in Arabic, which is often neither their mother tongue nor their second language.

The Organization of Islamic Cooperation (formerly Organization of the Islamic Conference), an unsuspected source, calculated that its more than 50 members have on average of only 8.5 scientists, engineers and technicians per 1000 inhabitants, whereas the average is 139.3 for the countries of the OECD, the Organization for Economic Co-operation and Development. The number of mathematicians among these 8.5 scientists is negligible. In addition, the number of scientific publications confirms this distressing situation: forty-six Muslim countries together provide only 1.17% of the publications, and twenty Arab countries together 0.55%. The American National Science Foundation observed that half of the 28 countries with the lowest scores in scientific articles belong to the Organization of Islamic Cooperation. Still, it is hard to imagine that Arab universities with renowned names like 'King Faisal University' or 'Prince Sultan University' would lack funding.

Moreover, some of those 'scientists' spend some of their time calculating the temperature of hell, or the chemical composition of a jinn, a spirit, like the one in Aladdin's lamp. No surprise, someone figured out that Mecca is in accordance with the golden section, that is, the ratio $\phi = 1.618\ldots$ On a world map, it lays exactly on the spot where two golden section rectangles intersect (Fig. 20.2).

It also seems this link between the location of Mecca and the golden section is emphasized in the Quran. The 96th verse of the chapter called 'The Family of Imran' (or 'The House of 'Imrán'), the third part of the Quran, can be translated as follows: 'Indeed, the first House [of worship] established for mankind was that at Makkah—blessed and a guidance for the worlds.' In the Arabic version, this verse has 47 symbols, and the word Mecca is in 29th place: $47/29 = 1.620689\ldots \approx 1.618 \approx 1.618\ldots$, the golden ratio. Clever! (Fig. 20.3).

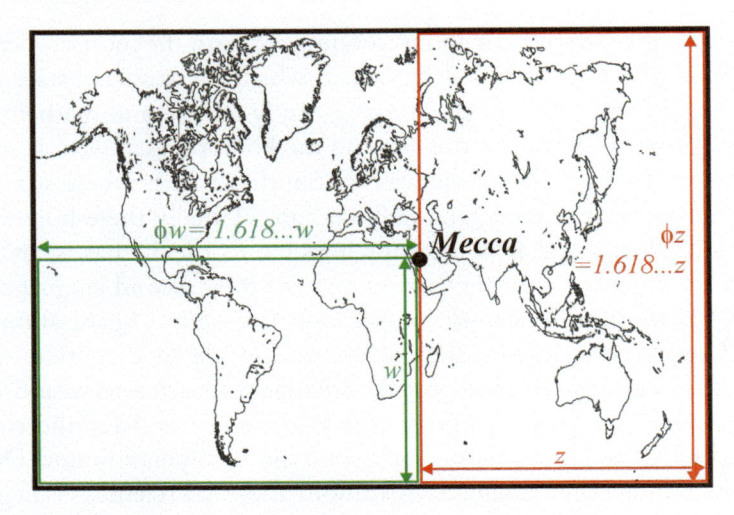

Fig. 20.2 Mecca's location in accordance with the golden section

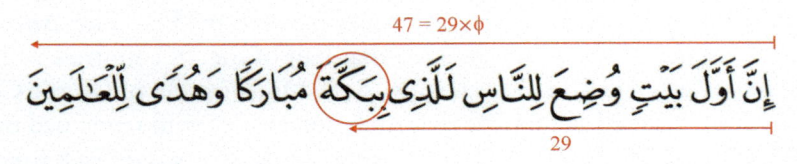

$47 = 29 \times \phi$

إِنَّ أَوَّلَ بَيْتٍ وُضِعَ لِلنَّاسِ لَلَّذِي بِبَكَّةَ مُبَارَكًا وَهُدًى لِلْعَالَمِينَ

29

Fig. 20.3 Mecca in the Quran at a location given by the golden section

A Great Past

Yet, it was Muslim scholars who dominated the mathematical world for nearly 700 years. In the 8th century there was Musa al-Khwarizmi who, in his House of Wisdom in Baghdad, compiled not only astronomical tables but also the oldest textbook in arithmetic, which would be used in its Latin translation until the 16th century in Europe. Incidentally, note that the word 'algorithm' is derived from his name. Around the turn of the millennium al-Biruni wrote many books, including about 15 works on mathematics, while he was living in what is now Uzbekistan. Sums of series, algebraic equations and the further development of Archimedes' theorems all caught his attention. In the 11th century lived, perhaps, the greatest mathematician, poet and philosopher of the entire Middle East, Persian Omar Khayyam. He realized that proportions in geometric figures do not necessarily lead to numbers that can always be written as fractions. He also had the idea of combining al-gebra (another word with an Arabic etymology) and geometry to solve equations of the form $x^3 + px^2 + qx + r = 0$. One of his contemporaries, Thabit Ibn Qurra, was the first to question the parallel postulate of Euclid, stating the intuitive fact that exactly one parallel to a given line passes through a given point. He studied the question in the 9th century, long before Nikolai Lobachevsky (Russia) or Janos Bolyai (Hungary) did in the 19th century (Fig. 20.4).

A mathematical 'Prince of Persia' was Ulugh-Beg (c.1393–1449), a grandson of Timur Lenk (1336–1405), who was first governor of Samarkand and then prince of the entire empire. He attached great importance to science and built an enormous astronomical observatory at Samarkand, a town in today's Uzbekistan. Under his rule, Quran schools became a kind of Islamic academy, where mathematics and astronomy were held in high regard. Between 1408 and 1437, some 70 astronomers worked there, including Al-Kashi (1380–1429). Although astronomy was the most important field of study, Al-Kashi would, for example, also calculate the decimals of pi based on an 8,050,306,368 side polyhedron. His value for pi was 3.14159265358979325, which is correct to the 16th decimal place (the 17th should be 4, not 5). His 'totally useless' world record would stand until 1596 when the Dutchman Ludolph van Ceulen (1540–1610) calculated 20 decimals. Al-Kashi also calculated the sine of 1° with great accuracy, 200 years before Kepler. In 1449, Ulugh-Beg was murdered on his way to Mecca by

Fig. 20.4 Al-Biruni in the Persian Scholar Pavilion at the Vienna International Centre

Fig. 20.5 Ulugh-Beg, surrounded by his astronomers, and the remains of his observatory in Samarkand (Uzbekistan)

fundamentalists led by his own son. Perhaps the courtiers could no longer tolerate that he paid more attention to scientific than secular matters. His death brought an end to scientific activities in Samarkand (Fig. 20.5).

Mathematics Becomes Meaningless…

Ulugh-Beg's mathematical activities represented the pinnacle of Islamic mathematics. Towards the end of the 15th century, the influence of Islamic scholar Shaykh Ahmad Sirhindi (1564–1624) would have far-reaching consequences. This Punjabi jurist regarded mathematicians as 'idiots' and their admirers as 'worse idiots and the worst creatures'. In his Maktubat I/266 he wrote that mathematics is 'totally meaningless and absolutely useless.' After all, mathematics could not be of service to a man on his way to the afterlife. According to him, it sufficed to know how to determine the distribution of an inheritance and how to find the direction to Mecca for prayer. Since then, this view prevailed among the 'Ulama', the scholars who 'officially' study the view of Islam and Sharia.

Remarkably, almost at the same time Europe put an end to its long tradition of looking down on mathematics. It is true that Galileo Galilei

(1564–1642) was still in trouble for his scientific inventions, but from about this same period Western Europe gradually got rid of the yoke of religious obscurantism. Since Saint Augustine, many Christians had rejected science-for-science's sake, in a late-antique tradition, deeming the study of mathematics meaningful only in so far as it could contribute to 'true happiness.' For a long time, the 'negative numbers' were considered evil by the Catholic Church, which implied that only the Pope was allowed to study math, in his room, and not ordinary Christians. Times changed in the West and eventually the Jesuit schools, for example, would hold mathematics in high regard. It is as if around 1600 the paths of the West and the Middle East crossed, taking opposite directions, from opposite points of departure.

…And Meaningful Again

But it is not all doom and gloom. Firstly, it must be said that some European 'scholars' spent their time with meaningless science as well, as proven by the success of the 'Ig Nobel Prize', the satirical counterpart of the Nobel Prize. And conferences on math and art or architecture still feature golden section research like the one quoted above on the location of Mecca.

Secondly, the OECD statistics around research quality are subject to interpretation. Perhaps certain Western countries are getting good track records in the Quotation Index and other statistical methods measuring the importance of research, because math geniuses from the Middle East emigrate and work in Western countries. It isn't just the US born and raised scientists that elevate so many US math departments to the top of the world ranking. In 2014, Maryam Mirzakhani (1977–2017), for instance, was professor at Stanford University when she got the Fields Medal, the most prestigious award in mathematics, but she got her Bachelor of Science in mathematics in 1999 from the Sharif University of Technology in Tehran, Iran. This education must have been excellent as it led her to Harvard where she got her Ph.D. And she surely is not the only top Muslim mathematician, though she remains one of just two women to have received the Fields Medal, regardless of background or culture (Fig. 20.6).

Not wearing a headscarf, eating pork, or drinking wine was forbidden at the same time as mathematics was, but for some reason the malediction on mathematics faded away in the middle of the 20th century when many Muslim countries modernized and needed mathematics for their development. After all, mathematics does not care about religion or nationality.

Fig. 20.6 Reza Sarhangi, Iranian born promotor of mathematics

And so, the opposition to mathematics by Nigerian Islamists, fundamentalist Asian madrassas or Western Muslim hardliners may be a rearguard action, at least according to Reza Sarhangi (1952–2016), professor in the Department of Mathematics at Towson University (USA). He founded the conference series 'Bridges: Mathematical Connections in Art, Music, and Science' and, being of Iranian descent, he liked to give lectures on 'Islamic mathematics'. Over a glass of wine—he would proudly emphasize that it was Persia who invented this 'drink of the gods'—he couldn't stop giving reasons why 'math is beautiful; scientists have become too important, and no one gets the mathematical djinn back in the lamp, even if it's Aladdin's.'

21

The Church's Perspective

After the discovery of perspective during the Renaissance, its rules became so familiar that earlier painters and their techniques were looked down upon and even got derogatory denominations, such as 'the Flemish Primitives' or 'the reverse perspective'. Yet their drawing methods can also be explained according to a logical structure—it is not totally wrong

Fig. 21.1 A triptych by Jan van Eyck (Dresden 1437) and the 'Representation of the Virgin in the Temple' (Studenica, Serbia, 1314) neither seems to follow the usual rules of perspective

Perspective and Parallel Projection

Classical perspective rules in figurative painting today, all over the world. The perspective with one, two or three vanishing points was established by the contributions of famous Italians such as Filippo Brunelleschi (1377–1446), who gave a demonstration with a perspective drawing on the square in front of the Duomo in Florence, in 1413, and Leon Battista Alberti (1404–1472), who wrote the treatise *De pictura* (or *Della pittura*) on perspective drawing. In a classic explanation, the drawing of a cube, for instance, can be explained very simply by light rays reaching the eye from each corner of the cube and extending them onto the drawing surface, thus providing the perspective view. We can hardly imagine why it took so long for artists to apply it. Mathematicians too, because for a long time the study of perspective was a subject of predilection in both art and mathematics (Figs. 21.1 and 21.2).

The Last Supper by Leonardo da Vinci (1452–1519) is a well-known illustration of the one-point perspective (Leonardo finished it in 1498, but it was totally lost by the year 1556, and was thus overpainted repeatedly by others) (Fig. 21.3).

Parallel projection can be explained in a similar way as perspective with vanishing points. For example, let's look from a direction that makes an angle of 19.8° downwards from a horizontal plane and an angle of 19.8° to the left from a vertical plane directly in front of the observer (a so-called profile plane). If a cube is on a horizontal plane such that the face in front is frontal

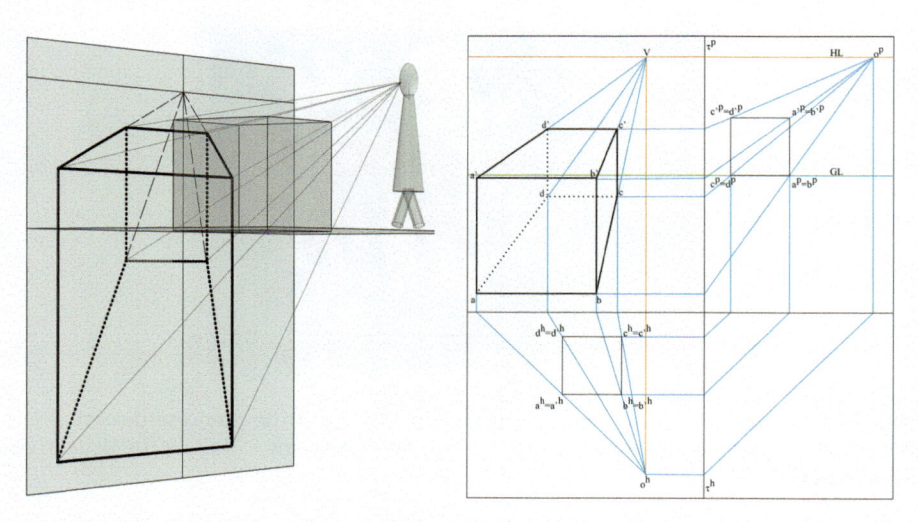

Fig. 21.2 Traditional construction of a perspective drawing with one vanishing point, in a 3D view and according to the Monge method

Fig. 21.3 The *last supper* by Leonardo da Vinci

to the observer (and thus also the face at the back), then the lengths of the edges of the faces in the final drawing have either the same length as the other edges or their half. The latter is the case with the edges that seem to go backwards. They make an angle of 45° with the horizon. This is the so-called cavalier projection, a very simple representation that follows from the choice of those angles of 19.8°. A choice of other angles instead of 19.8° yields projections that are not cavalier projections, but, for instance, so-called *isometric* or *dimetric* representations.

Many textbooks, as well as Wikipedia, add that 'the objects are not in perspective, so they do not correspond to a representation of an object as it can be seen in reality, although that drawing method does produce somewhat convincing and useful images.' This seems exaggerated, because the parallel projection can be interpreted as a perspective view with parallel sightlines, creating vanishing points 'at infinity'. With some ingenuity one can imagine that this is the way in which someone with dreamy eyes perceives reality or someone who is very 'far away'. Therefore, in the diagrams given here, an angel was used to represent the observer (Fig. 21.4).

So not only is the parallel projection not an erroneous method, it is also very easy to perform. After all, frontal shapes can be represented in their actual form, while in the cavalier representation, for example, the vanishing

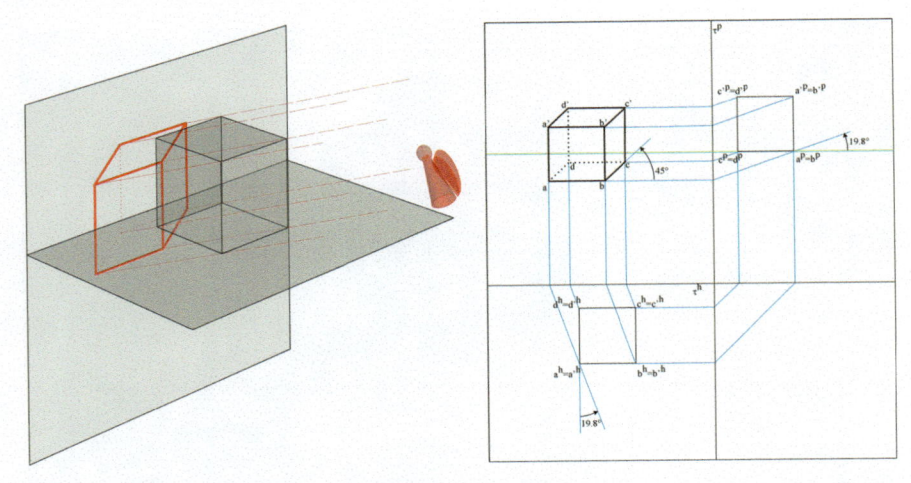

Fig. 21.4 Explanation of the cavalier construction: 3D view with an angel looking along parallel lines, and a similar construction using Monge's method

lines are parallel to the diagonals of a frontal square and the distances only have to be shortened by half. It is therefore not surprising that shops such as Ikea used the drawing method in their very first furniture montage manuals, and that many computer games used its simple graphical construction to generate two-dimensional images of three-dimensional objects. In Asia too, it was also widely used in traditional paintings (Fig. 21.5).

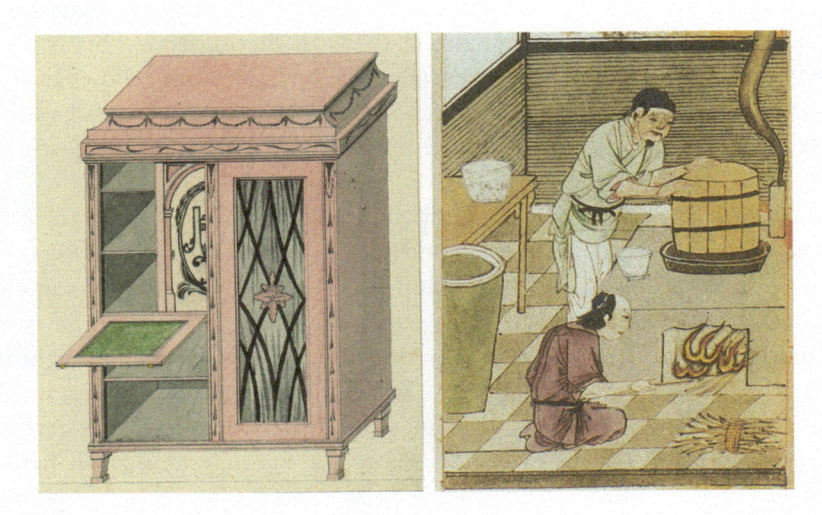

Fig. 21.5 Drawings in cavalier representation: part of *Drawing, Designs for Two Cabinets* (c.1805, Cooper Hewitt, Smithsonian Design Museum) and a part of a Chinese picture

A Mixed Form: The Fishbone Representation

The classical Renaissance perspective gained such importance that masters of that technique looked down upon earlier painters or artists from other cultures who did not adhere to the rules. For example, the term 'Flemish Primitives' refers to the 'primitive' methods in the paintings of the fifteenth and sixteenth centuries from the then Burgundian and Habsburg parts of what are now Flanders and the Netherlands. Initially it was a derogatory term, but over time it was reappropriated as an honorary title, a kind of badge of honor. The term is now proudly propagated by the Flemish government in foreign publicity campaigns (and rightly so) (Fig. 21.6).

Typically, lines of sight that should represent parallels do not converge to an identical central vanishing point, but to a vanishing line. German-American art historian Erwin Panofsky (1892–1968) coined the term 'fishbone perspective' and extensively discussed its symbolic interpretation. Its philosophical and art-historical context has been and still is described extensively in the literature, but it is less known that the representation can also be constructed mathematically. Here we focus on this technical aspect, that is, on how to draw this 'fishbone perspective' according to the above-mentioned classical and simple methods (Fig. 21.7).

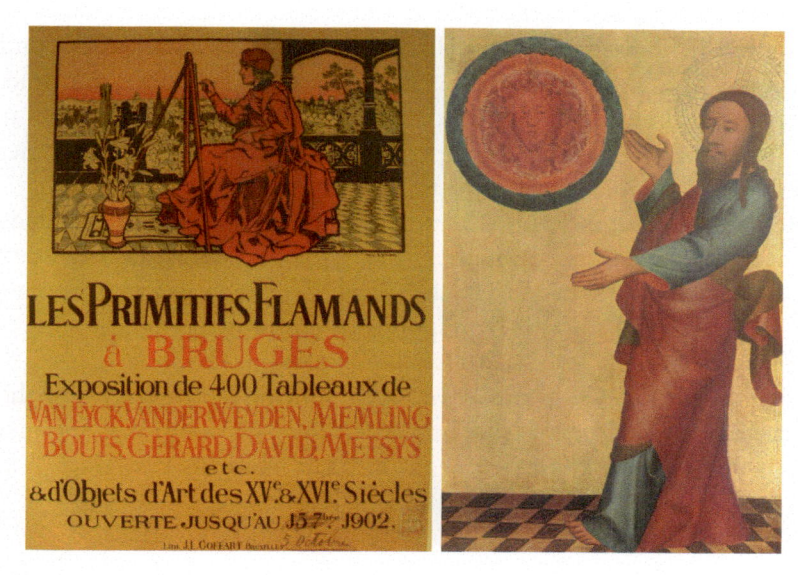

Fig. 21.6 In 1902, an exhibition used the term Flemish Primitives, which is somewhat understandable when considering the awkward covering of floor tiles (Grabow's altarpiece, by the German painter Master Bertram, c.1345–c.1415)

Fig. 21.7 Jan Van Eyck's *Dresden* triptych has a fishbone perspective

While teaching this subject, I noticed that the combination of both above methods naturally produces a fishbone representation. It suffices to draw the usual point perspective method in the top view and the usual parallel cavalier method in the left-side view. In three dimensions, this corresponds to the construction of the drawing in parallel layers one above the other. In each layer one draws according to the rules of perspective, and moves up or down, layer by layer. It is as if one performs the perspective drawing method correctly sitting higher or lower while drawing (Fig. 21.8).

One could, of course, also mix the perspective and cavalier methods the other way around, drawing the usual point perspective in the left-side view and the parallel cavalier method in the top view. That creates a horizontal fishbone. In three dimensions, it corresponds to the construction of the drawing according to parallel layers next to each other. In each layer one draws according to the rules of perspective, while moving layer by layer to the left or to the right. It is thus as if one carries out the perspective drawing method correctly, sitting more to the left or to the right while drawing (Fig. 21.9).

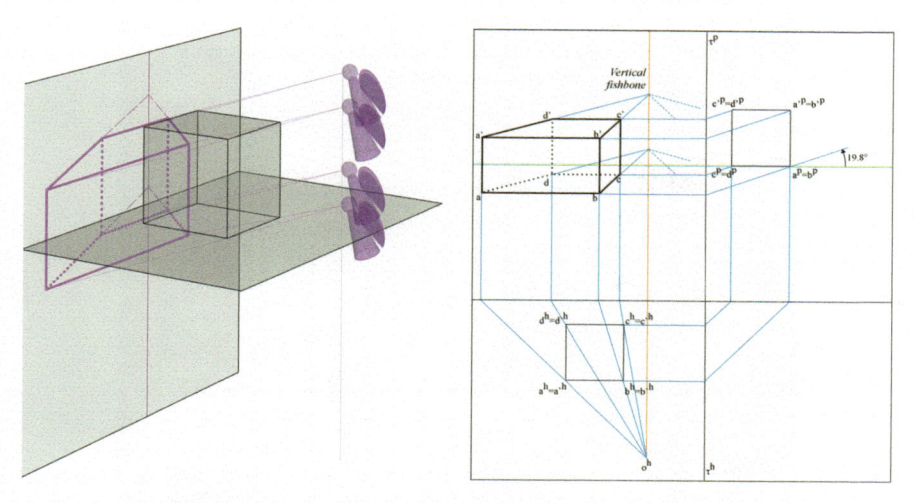

Fig. 21.8 Vertical fishbone construction: 3D view with an angel looking at parallel planes above or below each other and the same construction using the Monge method

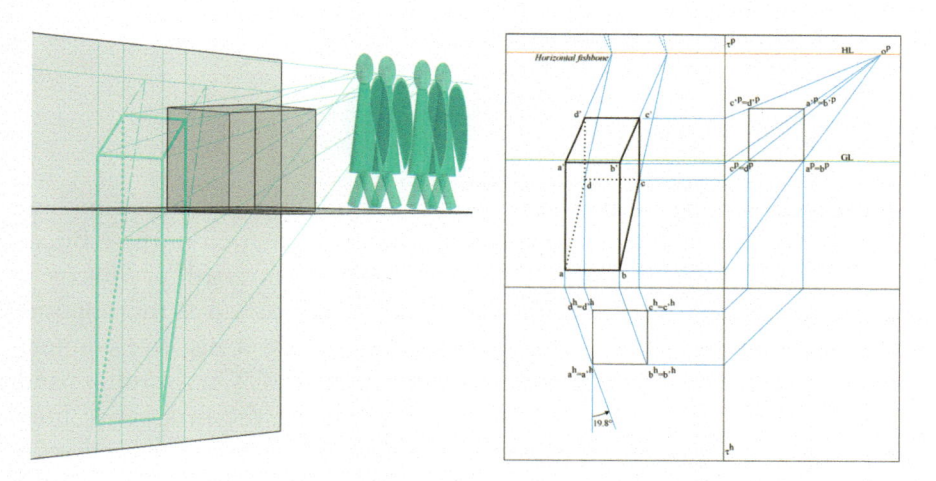

Fig. 21.9 Horizontal fishbone construction: 3D view with an angel looking along parallel adjacent planes and the same construction using the Monge method

These constructions show that the 'fishbone perspective' is a logical result of an algorithmic construction, which can be done on computer if desired. It is therefore more than the result of an artistic or philosophical idea. They also illustrate the analogy between the vertical and horizontal herringbone method (Fig. 21.10).

Fig. 21.10 Example of a horizontal fishbone: *The Flagellation of Christ* (c.1320) by Pietro Lorenzetti

Reverse Perspective

For certain iconographic art, mainly from Eastern Europe, the above methods do not provide an adequate explanation. The representation in these Byzantine drawings sometimes bears the telling name of 'inverted' or 'divergent' perspective. On the one hand, the name can be interpreted as the 'opposite' of the 'normal' perspective, and could thus also have a negative connotation, as was the case with the name 'Flemish Primitives'. On the other hand, it can also refer to the apparently opposite sense of the perspective lines, which seem to approach the viewer of the drawing instead of moving away from him. The usual geometrical methods applied above can still explain this method though, by placing the draftsman and the observer on opposite sides of the projection scene. It doesn't change the construction itself much, but the visible and hidden lines do swap (Fig. 21.11).

Some authors, such as the already quoted Panofsky see this reverse perspective as a method deliberately creating the impression that the drawing comes from heaven and descends to man. It would increase the heavenly impression of the icons (Fig. 21.12).

21 The Church's Perspective   199

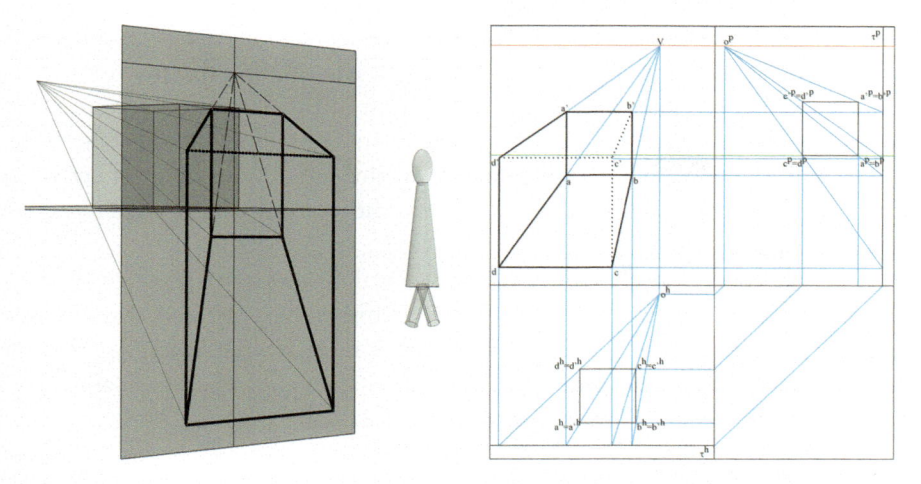

Fig. 21.11 3D construction and a similar Monge construction for the reverse perspective: the artist stands behind the screen, the viewer in front of it

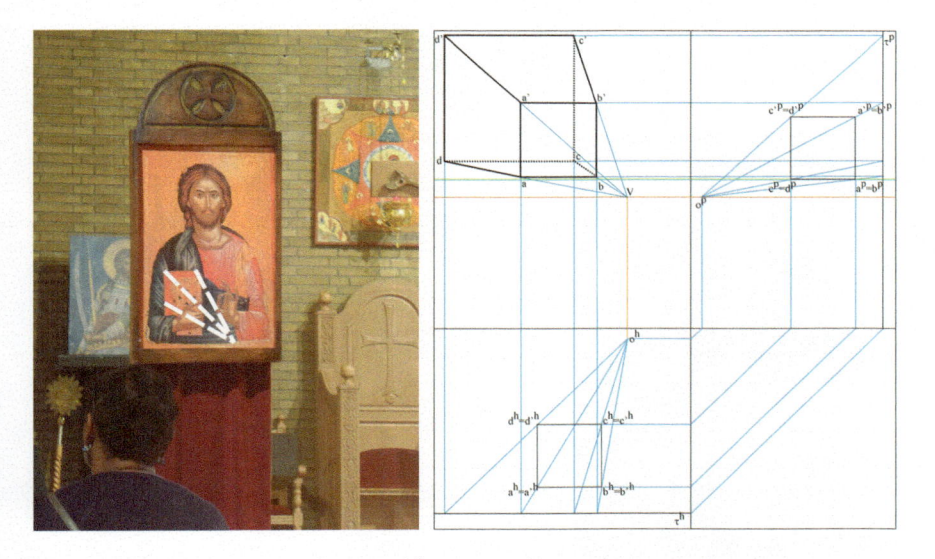

Fig. 21.12 Reverse perspective would create a heavenly impression

The fishbone representation and the reverse perspective can be combined too. It produces what might be called the 'reverse horizontal or vertical fishbone' perspective. The algorithmic aspect of these construction methods also explains why there are so many intermediate forms. Admittedly, some critics interpret this differently, asserting that the artists did not really know what they were doing. Some argue that the reverse perspective is a myth, an interpretation a posteriori. Perhaps this is too negative, as artists did experiment with different methods. Here we used the 'modern' 3D representations and Monge's method to explain them, but an experienced artist did not need these schemes and could have based his drawings on intuition and observation. He could have constructed images in vertical layers, moving from one side to the other, while making his drawing following the usual rules of perspective in every plane of view. This creates a 'reverse horizontal fishbone' perspective (Fig. 21.13).

There are also examples of inverted vertical fishbone perspective. It then appears as if the artist used oblique layers and moved up and down while creating his image, using a perfect perspective in each skew layer. Apart from philosophical considerations, this Byzantine art could therefore also have been inspired by logical, practical ideas (Fig. 21.14).

To this day, some artists use the classical rules of perspective in a creative way. Both the Belgian Jean Brusselmans (1884–1953) and the Frenchman Paul Cézanne (1839–1906) deliberately disobeyed the rules of classical perspective, and they are certainly not the only ones. Overwhelmed by the many perfectly drawn computer representations of 3D reality, there even seems to be a resurgence of all kinds of 'wrong perspectives' (Fig. 21.15).

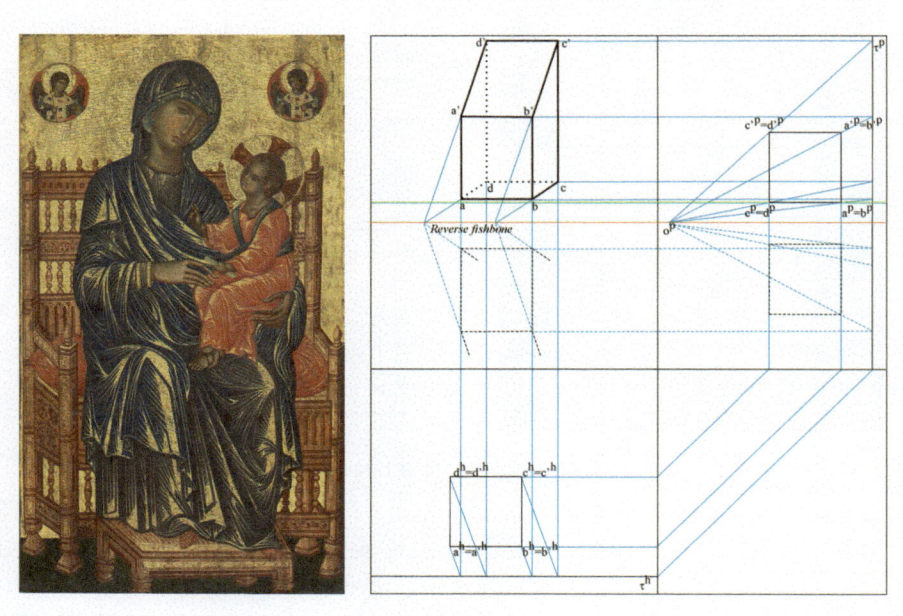

Fig. 21.13 A 'reverse horizontal fishbone' perspective with the corresponding Monge construction

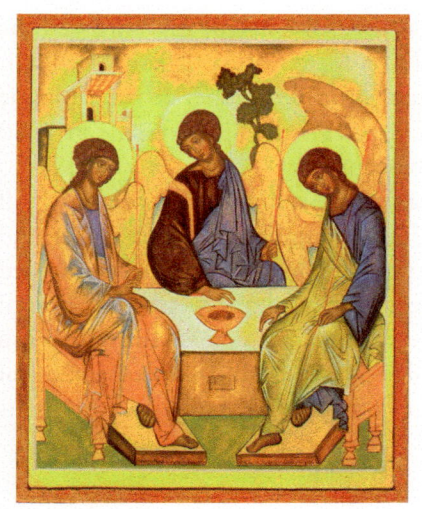

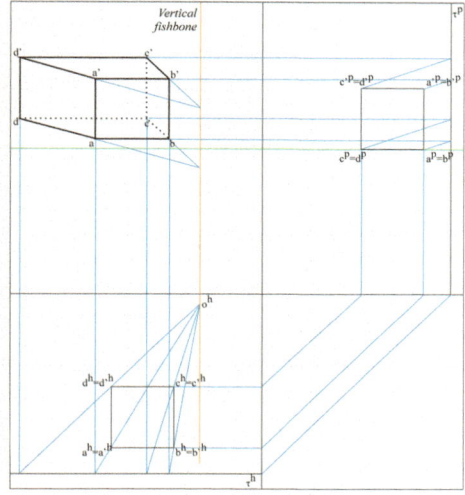

Fig. 21.14 Work by Andrei Rublev (Russia, c.1360–1428) using reverse perspective, which would create a heavenly impression

Fig. 21.15 Jean Brusselmans' *Woman in the kitchen* (1935, MSK Gent) and Paul Cézanne's *Still Life with Bottle of Liqueur* (1888–90, private collection, New York, VS)

Mathematical Saints

To conclude this chapter about ecclesiastical matters, here is some information about whom mathematicians could pray to. First, there is Albertus Magnus, also known as Albertus the Great or Albert of Cologne. He was born in Lauingen (Germany) between 1193 and 1206. He studied in Padua, Bologna, Regensburg, Freiburg, Strasbourg, and Hildesheim. He received a

'Ph.D.' at the 'University of Paris', which at that time meant that he had done in-depth studies at the monastery of the Dominicans of Saint James in Paris. In 1248 he organized the first Dominican 'Studium Generale' in Cologne (Germany), together with Thomas van Aquino. He died in Cologne in 1280 and was given a day in his honor on 15 November.

Albertus got the title of 'Magnus' because of his knowledge of almost all known fields in his time, and because of his reputation as a chronicler and critic of Aristotle. He favored what we would now call the method based on observation. Today's scientists may be disappointed to learn that one of his most significant achievements was the discovery that the Milky Way is a collection of stars. He also studied the solar eclipse during the crucifixion of Jesus. In mathematics his most important contributions were his comments on Euclid's *Elements*, such as:

> 'Any quantity remains a quantity when doubled, but some angles do not remain an angle when doubled, such as a right angle. For this angle, when doubled, does not remain a quantity. Therefore, an angle, by its very nature, does not appear to be a quantity.'

Albertus Magnus's observations must be seen in its historical context, when reading Euclid still seemed like a waste of time for a religious person, since all knowledge could, of course, be found in the Bible. Pointing out errors in Aristotle's work, as Albertus did, was courageous too because even Aristotle long remained a kind of saint in scientific circles. So, it is not surprising that many schools, universities, and institutes have been named after Albertus Magnus—and even a plant and an asteroid.

Albertus has a competitor for the title of patron of mathematicians: Saint Hubertus (c.655–727 AD), born in Toulouse (France) or, according to other sources, in the Belgian Voeren region. In 708 AD, he became bishop of Liège (Belgium). He died in Tervuren (Belgium), receiving his patronal feast or name day on 3 November. He is better known as the saint of the hunters and therefore of the makers of precision instruments such as spectacle makers and metalworkers because weapons were considered instruments of precision. And so, because of this aspect of rigor, he is sometimes mentioned as a saint of the mathematicians, although he did not write a single mathematical text (Fig. 21.16).

There is even a third possibility: Luke the Evangelist. He was a Greek, born in Syria, who lived in the first century AD, having his name day on 18 October. Luke became not only the saint of surgeons and butchers (often mentioned together in antiquity) and of the medical profession in general, but also of students, regardless of their subject of study (Fig. 21.17).

Fig. 21.16 The statues of Albertus Magnus at the University of Cologne (Germany) and of Saint Hubertus at the Museum M in Leuven (Belgium)

Fig. 21.17 Lucas at an easel, beside him a bull, his symbol (from *Les Images De Tous Les Saincts et Saintes de L'Année*, Jacques Callot, 1636)

Fig. 21.18 The frontal view of *Our Lady of Philermos* does not show the face of Mary (left), but the view from a certain perspective does (right)

Luke is perhaps more famous as saint of the painters, as he is often depicted as a painter himself. He even left a work of art, *Our Lady of Philermos*, which now hangs in the Art Museum of Montenegro, in Cetinje, the former capital of the Balkan country. Carbon dating of the painting estimated it to be from 400 AD or even later, so that Lucas must have been very old when he painted it. The painting itself has a mysterious aspect too: Mary's face can only be seen from a certain perspective (Fig. 21.18).

In summary, pure mathematicians can pray to Albertus Magnus, while applied mathematicians should turn to Saint Hubert. Luke the Evangelist takes care of the more artistic mathematicians as well as the students, perhaps the largest group most often in need of divine assistance while practicing mathematics.

22

The Flower of Life

This chapter ventures into the shadowy hinterland between mathematics and 'New Age' ideas. Sounds senseless, because at first glance there can hardly be an overlap between mathematics and spirituality. Yet, inspiring it is, at least for some artists..

Fig. 22.1 Jan Detavernier's installation *Entrɸphi* has a symmetrical part in the water. It is invisible, but because of the reflection on the water the whole structure can be imagined anyway

New Age Math

New Age adepts like all kinds of patterns. One of their preferred ones is made by overlapping circles with the same radius, arranged around a central circle. Depending on the shape they see in it, they call it 'flower of life', 'seed', 'egg', 'tree' or 'fruit'. The term flower of life was first used by Drunvalo Melchizedek, in his book *The Ancient Secret of the Flower of Life* (1999), where the pattern illustrates the cover.

The pattern appears in different cultures, and has been used by many artists, from Leonardo da Vinci to Coldplay (Fig. 22.1). The ancient Greeks of Cyprus seem to have been first, as an 8th–seventh century BC bowl with a pattern of circular arcs in the middle was found in Idalion. However, they have good competition from the Assyrians of the now northern Iraqi city of Dur Sarrukin, where a flower of life was drawn as early as the seventh-sixth century BC. The Romans often used the pattern as decoration, while Islamic and Gothic artists depicted it as well. In more recent times, they can still be found in window decoration of churches, even in small parishes. In China, the overlapping circles were on spheres under a lion's claw at the entrance of important buildings.

Perhaps anyone who ever received a stationery set with rulers and compasses has played with such patterns of circles. Start with one circle and measure its radius six times from a given point on that circle. Draw six circles with the same radius as the initial circle around those points. That makes seven circles in total. When circles are added onto these six circles in the same way, starting from their intersections, that makes 12 more circles. In total, there will now be 19 circles. The construction can be expanded in the same way, to 37 circles and more. Their number obeys the formula $n^3-(n-1)^3$ (Fig. 22.2).

What kind of math can be done with this revered pattern? Sure, it aids in getting reference points to draw Plato's five regular polyhedra: the tetrahedron, the cube, the octahedron, the dodecahedron, and the icosahedron. Based on his readings of Euclid's books, Luca Pacioli wrote about this in his book *Divine Proportion*, for which he asked Leonardo da Vinci to make illustrations. The beauty of these created a kind of polyhedra-mania which lasts until today (Fig. 22.3).

Pacioli's divine proportion, that is, the number 1.618…, became known as the golden section and is now, as mentioned above, often noted as φ. The use of this Greek letter phi would be a tribute to Phidias, but alas, there is no proof he ever used this proportion. These and other misconceptions about the golden section are discussed in other parts of this book. Two

7	19	37

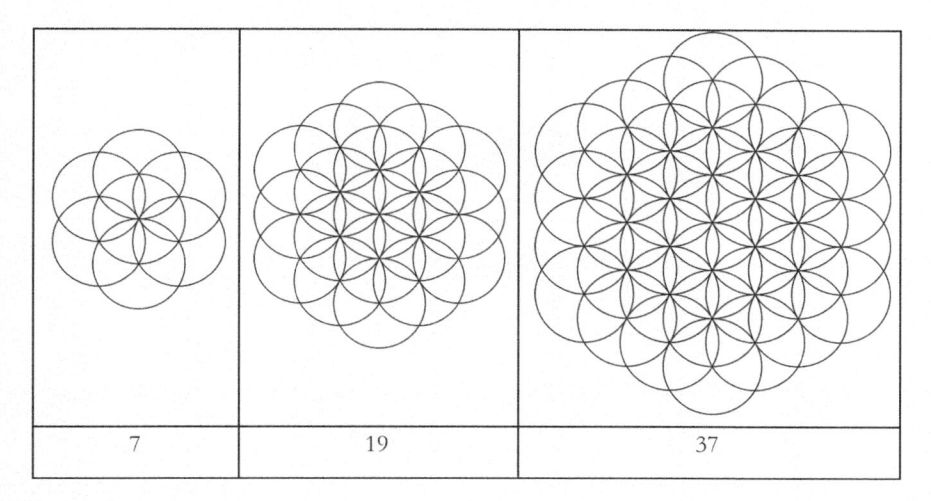

Fig. 22.2 Flowers of life

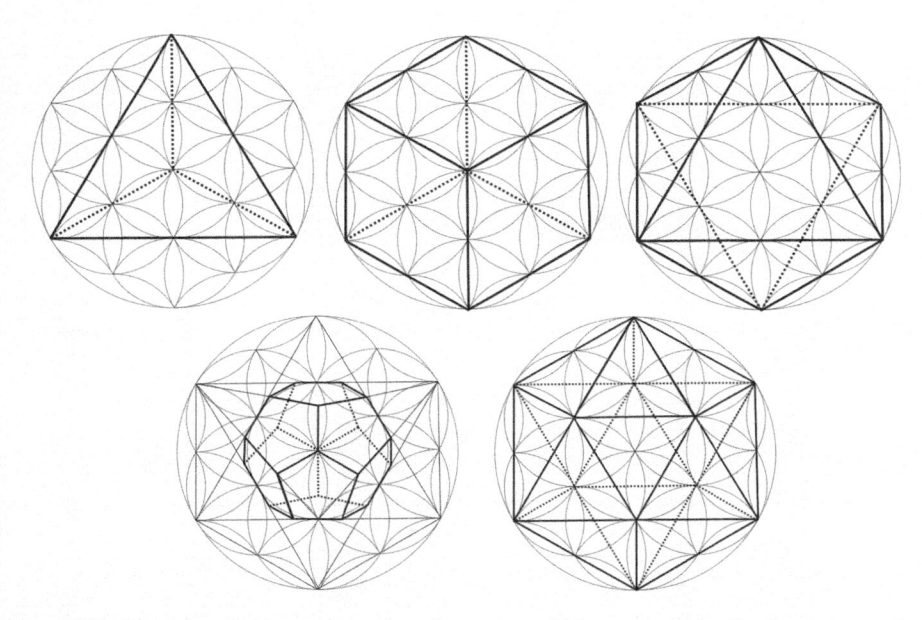

Fig. 22.3 The tetrahedron, the cube, the octa-, dodeca- and icosahedron drawn in a flower of life

figures given here illustrate some more funny New Age claims: sitting under a pyramid with a square base of side 2 and apothem ϕ would improve praying or studying; the symbol ϕ unites 'o', standing for 'the void', and I, 'the divine unity' (Fig. 22.4).

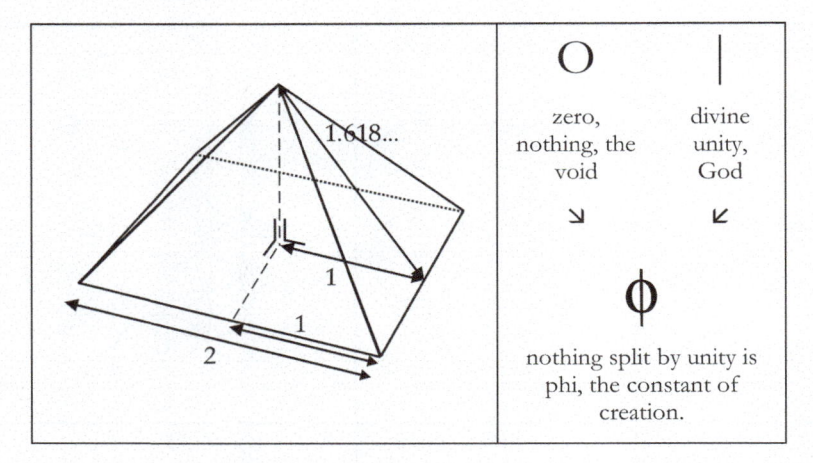

Fig. 22.4 The 'sacred pyramid' and the divine explanation for the golden section

Back to the flower of life, one can have some arithmetical fun too, by placing numbers from 1 to 48 in it, so that the sum of all numbers in certain circles and in certain parts of circles is always 294 (see the colored parts in the illustration). It is not a 'magic square' nor a 'magic cube', but a 'magic lotus', created by Indian mathematician Narayana Pandita (1325–1400). It appears in his work *Ganita Kaumudi*, or *Moonlight of Mathematics* (1356). This Indian origin fits well with the New Age mind set – and magic it is indeed (Fig. 22.5).

It is no higher mathematics, for sure, but mathematicians can feel honored that people turn to geometry or arithmetic in their spiritual quest, which

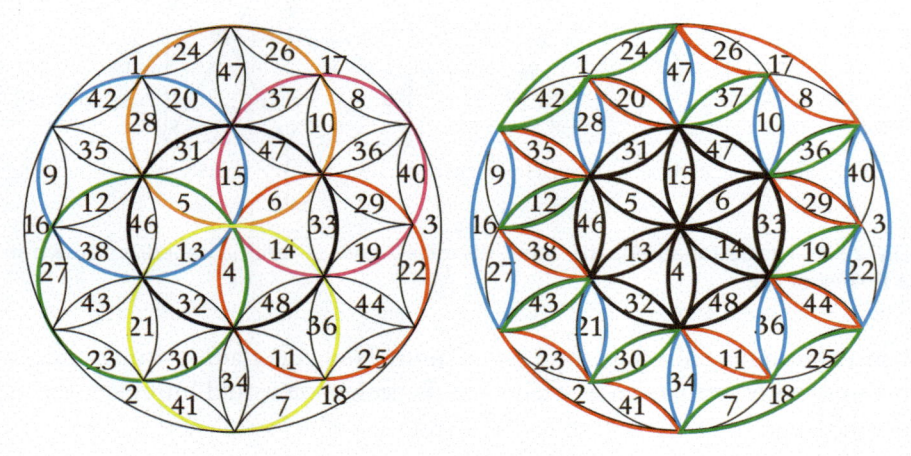

Fig. 22.5 Narayana Pandita's magic lotus: the numbers in the colored regions all add up to 294

Fig. 22.6 Jan Detavernier's installation *Dr*ϕ*ps*

is indeed common to many people. Belgian artist Jan Detavernier is so passionate about the flower of life, he made a drip machine that releases computer-controlled drops of water from a height on a flat surface of water so that the waves of the drops describe a flower of life. With some additional music and light effects, his installation *Dr*ϕ*ps* brings the viewer into 'the vibes of the Geometric Genesis', Detavernier explains, 'through a geometric ballet of circles' (Fig. 22.6).

Tetrahedra

New Agers also have a 3D approach to the flower of life, but that needs some more explanation. Begin with four tetrahedra and form a tetrahedron with them, placing three of them so that their edges align two by two and one of their vertices is common, again two by two, and put one of them on top of these three. Put four similar groups of tetrahedra upside down next to and on top of the first. In the remaining space fit three more. In the first step, there were 4 little tetrahedra, in the next step 20, and in the final one 32 (Fig. 22.7).

Now, start again with the group of four tetrahedra and form a larger one by placing three in a similar way next to each other and one group on the

Fig. 22.7 4, 20 and 32 tetrahedra

top. This is reminiscent of the so-called Sierpinski construction. Take another such large construction, turn it upside down and fit it in the first one. The first construction had 4 × 4 tetrahedra, and together with the 16 from the upside-down tetrahedron, that again gives 32 tetrahedra. Finally, this entire construction can be inserted in the previous 32 tetrahedra, so that a shape of 64 tetrahedra is obtained (Fig. 22.8).

Next, consider the top view, as if the initial tetrahedra stands on a horizontal plane. The vertices fit perfectly on the flower of life. Sure, some intersections of circles do not correspond to any point of the projection, but that does not bother a good New Ager (Fig. 22.9).

Covering some parts of this 64 tetrahedron structure with squares, one gets a cuboctahedron. It is an Archimedean solid made up of eight triangular faces and six square faces. Architect Buckminster Fuller liked it a lot, probably because he found a way to collapse the squares so that only the eight triangles remain and form an octahedron (Fig. 22.10).

In his typical jargon, Fuller also ascribed a 'toroidal flow' to the cuboctahedron. One can call it poetic or obscure language, depending on the admiration or aversion one has for these kinds of statements. In any case,

Fig. 22.8 16, 32 and 64 tetrahedra

Fig. 22.9 Top view of the 64 tetrahedra construction; a wire view; the same wire view on a flower of life

Fig. 22.10 The 64 tetrahedron structure, the same structure covered with additional squares, and the cuboctahedron

a deformed torus fits around the cuboctahedron, whatever that may mean or imply. New Agers see it as an 'energy flow' and call the center a 'singularity' (Fig. 22.11).

Moved by this poetic-geometric imagery, the abovementioned artist Jan Detavernier made an installation he provocatively called *Entrøphi*. The

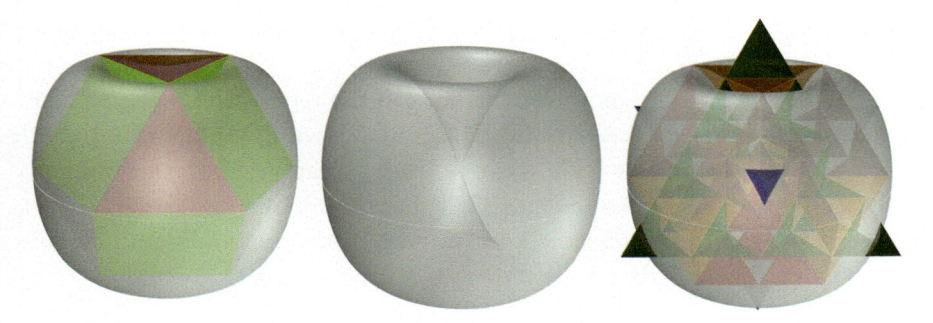

Fig. 22.11 The cuboctahedron with a deformed torus around it, the toroidal shape itself, and the original 64 tetrahedron structure with the deformed torus

Fig. 22.12 Jan Detavernier's tetrahedra with the toroidal flow

construction is a 3D model of the 64 tetrahedron structure. It has an invisible symmetric part in the water, but because of the reflection on the water the whole structure can be imagined anyway. To mimic the toroidal flow, he installed fountains on top of it. This artwork was exhibited in Menen, Belgium, in 2020 (Fig. 22.12).

Spatial Stacking

When I first saw the 64 tetrahedra construction, I was not greatly impressed by the toroidal flow or the link to the flower of life, but more by the procedure of stacking tetrahedra. Of course, stacking *cubes* on top of each other, in front of and behind each other, and pushing them together, fills the entire space without any gaps—just think of warehouses filled with cardboard boxes or beer crates. They may have some rectangular faces and thus not necessarily be cubes, but their faces are perpendicular to each other, and so they stack nicely, like cubes do.

Obviously, but what about tetrahedra? Do the tetrahedra in the 64 tetrahedra shape suggest one can fill space solely with tetrahedra? Aristotle (384–322 BC) thought so. He was one of the great Greek classical philosophers along with Socrates and Plato. Called the first 'homo universalis', he was interested in almost all sciences known in his time, including mathematics. In his work *De Caelo* or *On the Heavens*, Aristotle stated:

'Of all polygons it is known that there are three figures that fill the plane in which they lie: the triangle, the square and the hexagon; and of all spatial bodies only two, the tetrahedron and the cube.'

For 1800 years, commentators of Aristotle's text would go to great lengths to prove the great Greek scholar was right, until German Regiomontanus (1436–1476) realized that when stacking tetrahedra, there will always be some space left over, regardless of how one stacks them. After all, the angle between two faces is approximately 70°31.7…' and if you place 5 of them around a common edge, approximately 360°− 5 × 70°31.7…' = 7°21' remains. That is a small angle, which can hardly be noticed when making tetrahedra in stone or wood. Perhaps Aristotle had based his conclusion on 'empirical' observation, but theoretically it is not correct (Fig. 22.13).

So, what about the 64 tetrahedra structure? A closer look reveals that there are octahedral spaces between certain tetrahedra. And together, tetrahedra and octahedra can indeed fill space (Fig. 22.14).

While one shape of a prismatic brick is sufficient to build walls, two shapes should be used if one wants to involve tetrahedra. Regardless of the sliding effects that could occur when putting a tetrahedron on an octahedron, it does seem interesting and certainly novel to propose a tetra- and octahedral construction (Fig. 22.15).

Structurally, a tetrahedral shape surely is the strongest possible because it is only made of equilateral triangles, that cannot be deformed, whereas squares or rectangles can collapse easily. Will we soon see tetrahedral and octahedral bricks?

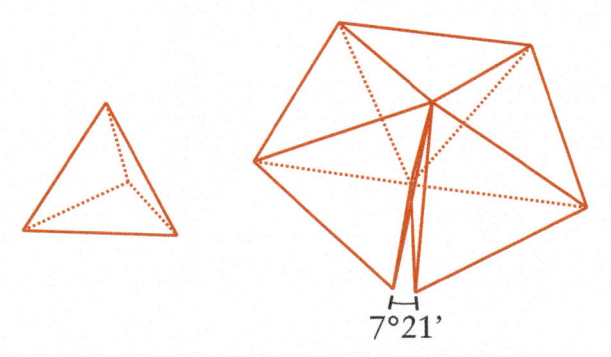

Fig. 22.13 One tetrahedron, and five of them, not filling space

Fig. 22.14 Removing the blue pyramid in front of the above 64 tetrahedron structure shows there is an octahedron hole in it (second and third column, below). Also, on each of the squares added to transform it in a cuboctahedron rest four pyramids that can be each be completed to an octahedron (second and third column, above)

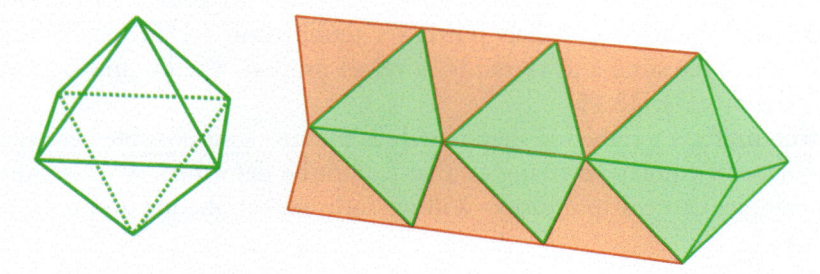

Fig. 22.15 An octahedron and a wall with tetra- and octahedral bricks

23

A Meta-Divine Nautilus

Circular arcs inscribed in squares whose sides are Fibonacci numbers will never approach the shape of a Nautilus shell, and yet prominent authors mention this as fact. There is no connection with the golden ratio 1.618… and perhaps neither with 1.333…, as others have also proposed. So, what is the real ratio of the Nautilus shell?

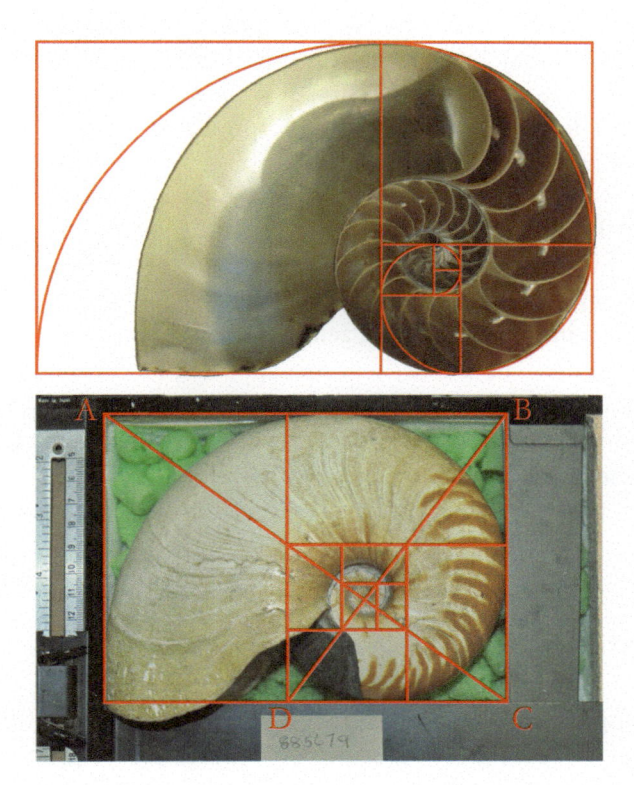

Fig. 23.1 Circular arcs in squares with sides 1, 1, 2, 3, 5, 8 and 13 never overlap a Nautilus shell nicely (top), but rectangles with a ratio of 1 to 1.355 do, at least, for a certain species (bottom)

A Beautiful Shell

Put a square with side 1 beside another square with side 1. Stick a square against them with side 2. Then fit a square with side 3 against them. And so on, until you get a series of adjacent squares with sides 5, 8, 13, and so on. Draw a quarter circle in each square and connect the arcs to each other: a spiral-like figure appears. The numbers 1, 1, 2, 3, 5, 8, 13, ... form the well-known 'Fibonacci sequence', named after Italian mathematician Fibonacci (or Leonardo of Pisa, 1170–1250). There is a fun property of the consecutive fractions formed by dividing each number by the previous one: $1/1 = 1$, $2/1 = 2$, $3/2 = 1.5$, $5/3 = 1.666...$, $8/5 = 1.6$, $13/8 = 1.625$, ... the numbers produced go up and down, and yet they reach the number $1.618...$, (e.g. $89/55 = 1.618...$) the so-called 'golden ratio'. That is why the 'Fibonacci spiral' described above, is also called the 'golden spiral'. This gracefulness of a spiral formed in squares may have been an inspiration for many to stick it on top of a photo of a cat sleeping in a curl, or on a model swirling her long wet hair, or on the hairstyle of a former U.S. president. These pictures are fun mathematical moments on social media.

It becomes more serious when the spiral is also placed over a Nautilus shell (Fig. 23.1). This shell has been admired as the embodiment of a graceful spiral shape for centuries. The chambered Nautilus is the sole cephalopod with an external shell, with the same body and shell arrangement as its ancestors. Still found in the deep Indo-Pacific, it can be considered a 'living fossil' from before the dinosaurs. The shells first appeared in Europe in the sixteenth century as rare exotic objects, which were richly polished or engraved for jars, cups, and table decorations, and decorated with bronze, copper, silver, and precious stones. They became status symbols in the cabinets of curiosities of the upper classes. Today they can be seen in museums all over the world (Fig. 23.2).

Shell Statistics

Not surprisingly, some wanted to make it even more beautiful, and make it correspond to the Fibonacci sequence. But that is simply not possible. Everyone can see that with the naked eye. And yet influential authors make this claim, such as British mathematician Keith Devlin (Stanford University, USA) in *The Myth That Will Not Go Away* (2007), Israeli American astrophysicist Mario Livio (Space Telescope Science Institute) in *The Golden Ratio* (2003), and, less surprisingly, writer Dan Brown in *The DaVinci Code*

Fig. 23.2 Nautilus shells in the Louvre

(2003). Mark Ryan pertinently illustrated the title of his book *Geometry for Dummies* (2016) with his claim that 'the spiral of the golden rectangle [...] has the same shape as the spiral shell of the Nautilus'. Even the eminent Princeton University Press put a Nautilus shell on the cover of the book *The Princeton Companion to Mathematics* (2008), edited by 1998 Fields Medal winner Timothy Gowers. On inquiry, the cover designers admitted having been inspired by the alleged role of the Fibonacci numbers in the shell, as shown in the opening picture.

However, as early as 1883 King's College Rev. Moseley (London, UK) noted that the Nautilus spiral grows by a factor of 1.316... over one quarter turn: the number 1.618... has nothing to do with it. More than a century later, he was joined by two mathematicians, John Sharp from Britain (2002, London Knowledge Lab) and the American Clement E. Falbo (2005, Sonoma State University), who stated: 'Anyone who has such a shell can immediately see that the ratio is somewhere around 4 to 3'. This gave rise to a new myth, namely that 1.333..., not 1.618... describes the growth of a chambered Nautilus, a claim that was proclaimed as truth on Wikipedia. Rachel Fletcher (1988) tried to save the golden ratio, by claiming that the factor is in fact 1.272 (and that is close to 1.333...), of which the square is, oh coincidence, 1.618.... In the online publication *Is the Nautilus shell spiral a golden spiral?* (2014), Gary B. Meisner adopted this point of view. He is known for his very popular book *The Golden Ratio—The Divine Beauty of Mathematics* (2018).

One problem with almost all of these claims is that they are based on the measurement of a single shell. As if one would substantiate claims about, for example, human qualities based on those of one individual. It turned out that Falbo, who claimed he had measured numerous shells from the California Academy of Sciences, also checked only a few shells from the Teaching Collection. Indeed, the curator of the collection could not find any official record of access in Falbo's name. Only recently, in 2018, someone took up the challenge of doing an actual statistical study.

Ch. Bartlett, a professor of the Art Department at Towson University (USA), would measure some 80 shells at the Smithsonian National Museum of Natural History in Washington D.C. There were six species: *Nautilus pompilius, repertus, belauensis, macromphalus, stenomphalus* and *scrobiculatus*. Proceeding systematically, he fixed the shells in a kind of easel and measured the increase of the 'spokes' from the center to the edge (see opening picture). The spoke from the center to A turned out to be on average 1.310 times longer than the one to B, for the 65 shells of the first five listed species, while this length turned out to be on average 1.310 times longer than the length of the spoke from the center to C, etc. For the 15 shells of the *Nautilus scrobiculatus*, the average was 1.355…. Thus, no trace of the golden ratio, nor of 4/3 = 1.333. The ruling numbers turned out to be 1.310… and 1.355… and yet it turned out there was something curious with these numbers.

A Meta-Divine Ratio

In his quest, Bartlett was motivated by a result that he had previously obtained 'by chance' (as far as chance exists in research). It was related to another myth: that of the golden ratio. It states that a rectangle with width 1 and length 1.618… would be viewed as the 'most beautiful rectangle'. From the Athenian Parthenon to the paintings of Leonardo da Vinci and to Mozart's music, all masterpieces in art would be witness to the aesthetic value of the golden ratio. It is a topic mentioned several times in this book. It is a myth created in the second half of the nineteenth century by German psychologist Adolf Zeising (1810–1876). Romanian diplomat Prince Matila Ghyka (1881–1965) popularized it so effectively that afterwards Le Corbusier (1887–1965), Salvador Dali (1904–1989), and in their footsteps many others, eagerly spread this 'fake news'.

As with most pseudoscientific stories, there is some partial truth in the golden ratio. Leonardo da Vinci indeed illustrated the book *De Divina Proportione* by Luca Pacioli, and he did so beautifully. The title of that book

23 A Meta-Divine Nautilus

is reminiscent of *La Divina Commedia* or *The Divine Comedy* by Dante Alighieri and was also the reason for that alternative name 'divine proportion'. The number itself does indeed have almost divine properties. For example, it can arise by connecting the center of a side of a square with side 1 with an opposite vertex and placing that connecting line on that side from that center (see illustration below).

The rectangle thus created has a length of $1 + 1/1.618... = 1.618....$ This immediately contains a nice numerical property: dividing 1 by $1.618...$ gives $0.618...$, with the same decimal digits, so that indeed $1 + 1/1.618... = 1 + 0.618... = 1.618....$ In a fun and correct way, this construction does indeed create a golden rectangle, that is, a rectangle with width 1 and length $1.618....$ But to say that this rectangle is therefore the most beautiful one is a bridge, or rather, a rectangle, too far (Fig. 23.3).

Moreover, this method of construction is not an exceptional property that would only apply to the golden ratio. On the contrary, if, in any rectangle with width 1, the center of the other side is connected to an opposite vertex, and if that connecting line is placed on that side from the center, this always creates a rectangle proportional to the enclosed smaller rectangle to be added to the original rectangle. That sounds complicated, but a drawing and a simple example make it evident immediately (Figs. 23.4 and 23.5).

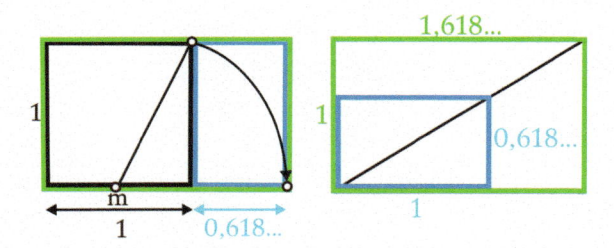

Fig. 23.3 A simple construction of the golden rectangle

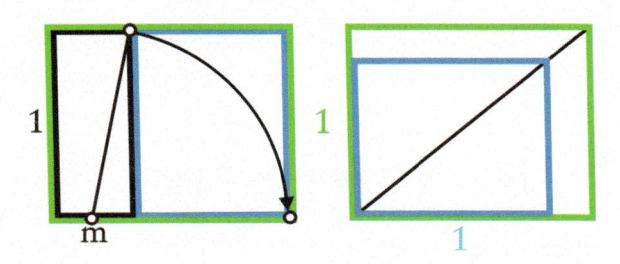

Fig. 23.4 Extension of an arbitrary black rectangle by a blue rectangle (left) such that the larger green whole has the same shape as the added blue rectangle (right)

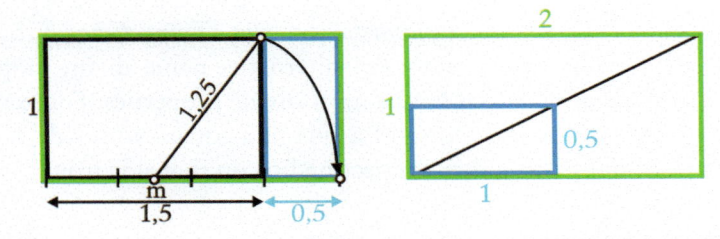

Fig. 23.5 A numerical example: the black rectangle has sides 1 and 1.5, so that the blue rectangle has sides 1 and 0.5 and the green sides 2 and 1 (right)

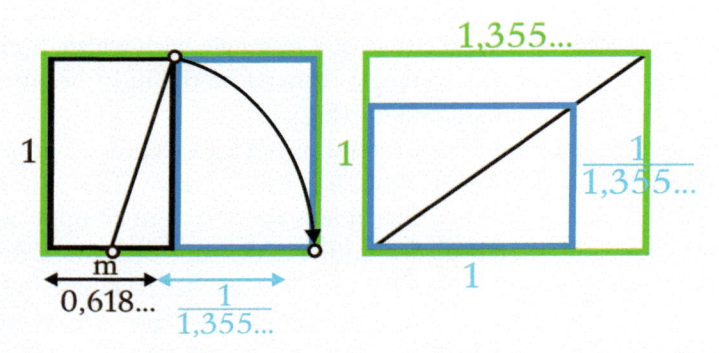

Fig. 23.6 The rectangle with sides 1 and 0.618... extended via the same method gives a rectangle with sides 1 and 1.355..., the meta-divine ratio

For example, take a rectangle with width 1 and length 1.5. The length of the connecting line of the middle with an opposite vertex then follows from the Pythagorean theorem: $1 + (1.5/2)^2 = 1 + (0.75)^2 = 1.5625$ and of this the square root is 1.25. Adding 0.75 creates a rectangle with width 1 and length 2. This is of course proportional to a rectangle with width 0.5 and length 1 because one only has to multiply both by 2.

Bartlett's idea now was to take a golden rectangle itself as starting point and expand it in a similar way. The obtained rectangle turned out to have width 1 and length 1.355.... He called this number the 'meta-divine ratio', with a nod to the divine ratio he had thus expanded (Fig. 23.6).

Similarities with the Divine Number

Mathematically speaking, this was a nice idea, because the number turned out to share many properties with the golden number. The golden or divine ratio is denoted by ϕ, but for sake of simplicity we use a separate symbol for $1/\phi = 0.618...$, namely ϕ'. The meta-divine ratio then gets the letter

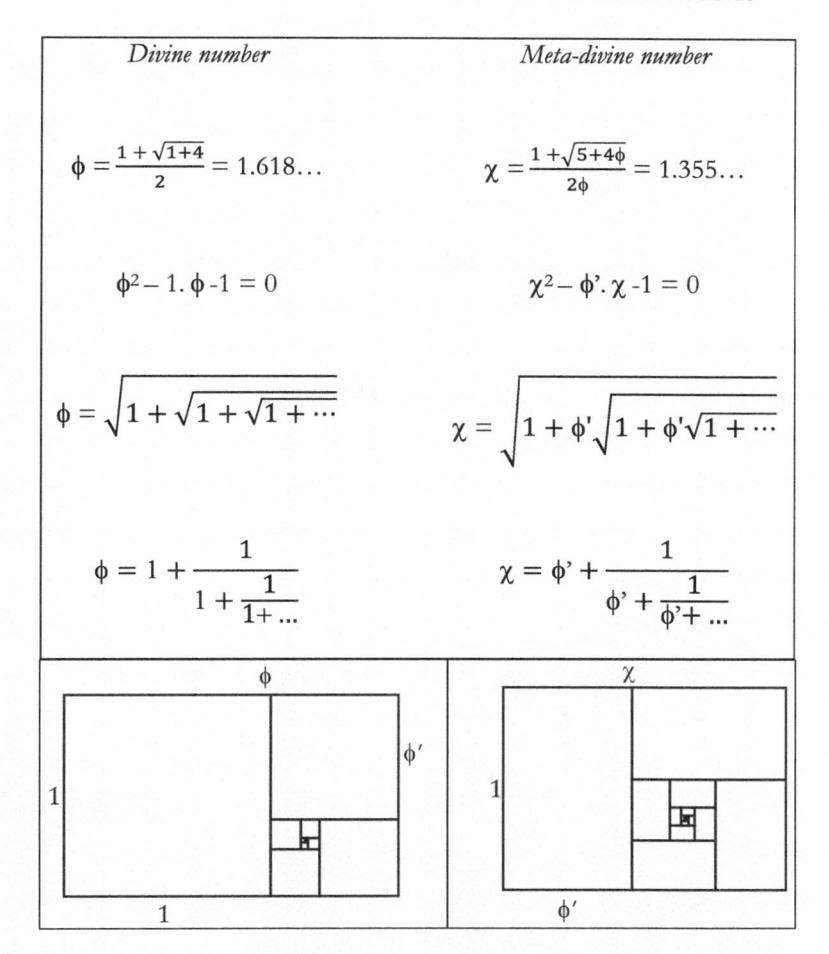

Divine number	Meta-divine number
$\phi = \dfrac{1+\sqrt{1+4}}{2} = 1.618\ldots$	$\chi = \dfrac{1+\sqrt{5+4\phi}}{2\phi} = 1.355\ldots$
$\phi^2 - 1 \cdot \phi - 1 = 0$	$\chi^2 - \phi' \cdot \chi - 1 = 0$
$\phi = \sqrt{1+\sqrt{1+\sqrt{1+\cdots}}}$	$\chi = \sqrt{1+\phi'\sqrt{1+\phi'\sqrt{1+\cdots}}}$
$\phi = 1 + \dfrac{1}{1+\dfrac{1}{1+\ldots}}$	$\chi = \phi' + \dfrac{1}{\phi' + \dfrac{1}{\phi' + \ldots}}$

Fig. 23.7 Table of similarities between the divine and meta-divine number

χ or chi because that is the next letter after φ in the Greek alphabet. From the following diagram, even the non-mathematician can visually comprehend that the numbers share beautiful properties. The math enthusiast can check them easily, if necessary, with a calculator (Fig. 23.7).

A New Myth?

Historically and artistically, the ratio turned out to be close to the so-called 'sesquitertia' ratio, which for centuries stood for 4/3 or 1.333.... It was one of the proportions Roman architect Vitruvius (85–20 BC) advised in his work *De Architectura* or *On Architecture*. This treatise with all kinds of rules of thumb was one of the few Roman manuals on construction that had survived the Middle Ages. The Renaissance brought the recommended ratios of 2/3 = 0.666... (the sesquialtera), 1/2 = 0.5 (the 'diapason'), and so on, back to life—again, there is no mention of the golden ratio here. Bartlett's chi ratio of 1.355... corresponded to this sesquitertia ratio, but he was not the first to propose a ratio close to 1.333.... In 1928, Dutch architect Dom Hans van der Laan had already introduced the plastic number psi $\Psi = 1.324...$, while in 1973 Spanish architect Rafael de la Hoz had proposed the Cordoba ratio $c = 1.306...$. These numbers do seem to play a role in the story of the Nautilus shell, because five species obey the 1.310 ratio, while the sixth, the *Nautilus scrobiculatus*, follows the meta-divine ratio, 1.355...

Incidentally, in 2015 there was an impressive study by Michael Trott, chief scientist of the famous Wolfram Research, which develops the top computer software Mathematica. He looked at an extensive dataset of more than a million paintings and had the software analyze the ratios of their canvases over the centuries. He concluded:

> 'The median of the proportions of paintings decreased over the past 500 years [...] The average also declined and seemed to stabilize at just over 1.35.'

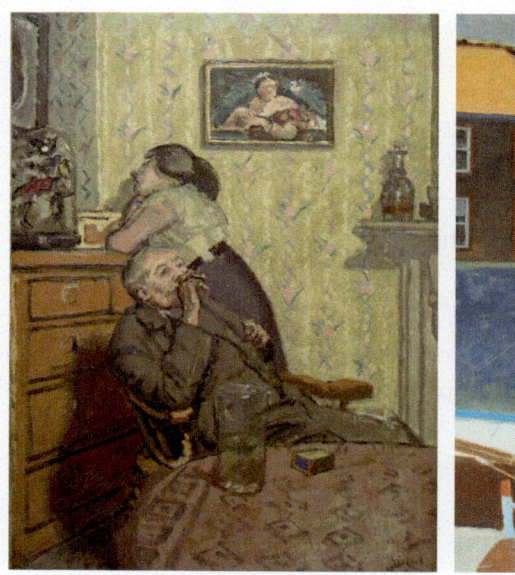

Fig. 23.8 Sickert's *Ennui* and Bartlett's *Cornish Coast*

In other words, over the centuries, it did turn out there was a 'preferred rectangle', namely one with a ratio of 1 to 1.35. An example is W. Sickert's *Ennui* (Tate Museum, London), which measures 152.4 cm by 112.4 cm, so that the proportion is indeed 1.355…. Inspired by this, Ch. Bartlett used the chi proportion in his own paintings, such as in his *Cornish Coast* (54 cm/ 40 cm = 1.35) (Fig. 23.8). Thus, it is not a rectangle with a ratio of 1 to 1.618… that turns out to be 'the most beautiful rectangle', as the myth of the golden ratio puts it, but one with a ratio of 1 to 1.35—and 'just over it', to quote Trott. Is that ratio 1.355?

24

Happiness in Unprovability

Harvey Friedman spent half a century on mathematical problems that cannot be answered. He made it his life's achievement to render this inaccessible mathematics tangible, occasionally writing a paper on 'A Divine Consistency Proof for Mathematics' as well. A conversation with a man who is as warmhearted as he is intelligent.

Fig. 24.1 Friedman: 'I'm working on a book with chess problems of a mathematical nature, but I don't want to share them yet. Sorry about that'

An Interview with Harvey Friedman

Harvey Friedman (1948–) is a genius: at the age of eighteen he obtained his doctorate in mathematics from the prestigious Massachusetts Institute of Technology. He was immediately appointed as an assistant professor of philosophy at Stanford University, thus earning the title of youngest professor ever in the Guinness Book of World Records. He then became professor at the University of Wisconsin at Madison and the State University of New York at Buffalo, before settling as professor of mathematics and music at Ohio State University, in 1977.

Three years earlier he had been a speaker at the quadrennial International Congress of Mathematicians, the most important mathematics congress in the world. In 1984, he won the prestigious Alan T. Waterman Award. The next year, a survey about his work appeared: *Harvey Friedman's Research on the Foundations of Mathematics*. In 2007 he won the Tarski Lectures Prize. Officially retired in 2012, he still won an honorary doctorate, in 2013, from Ghent University (Belgium) and so he decided to further strengthen his ties with that institution. In 2021 he became a guest professor in the research group in mathematical logic and set up a permanent archive of Ghent professor Andreas Weiermann (Fig. 24.1).

It was time for an interview, but how does one approach a man who specialized in mathematical logic and once wrote a paper entitled 'A Divine Consistency Proof for Mathematics'? Here is an excerpt from his paper:

> 'Following Gödel and going back at least to Leibniz a class of objects is positive if and only if it contains God. […] But then it would appear that one cannot expect to do anything substantial, mathematically. […] Interactions with scholars from theology […] rekindled our interest in trying again to do something powerful […]. It occurred to us that if we take God out of the class of all objects, treating God as exceptional, […] then the 'positivity ultrafilter' is no longer trivial.'

Moreover, Friedman's participations in lectures are famous for his quick thinking and well-founded questions, immediately, on the spot. Thus, I asked for support from Andreas Weiermann, who submitted the questions to Friedman. It generated some unrestrained straightforward answers.

About His Work

Q: Can you illustrate your research in a few sentences?

A: *In mathematics you ask questions to which you want unambiguous answers. You want to know if your claims are true or not. But in the 1930s, Austrian American Kurt Gödel (1906–1978) discovered there are mathematical hypotheses that you can neither prove nor disprove. They are totally unreachable. Gödel's work was rather abstract and far from mathematical reality. The desire grew to make his work more down to earth and relate it to the physical reality.*

It became my research topic for 54 years. I proposed tangible problems related to Gödel's more philosophical claims. I formulated everyday mathematical problems that cannot be answered within mathematics itself.

Q: Doesn't that make you depressed? Russian mathematician Vladimir Arnold once said Gödel's discovery had discouraged him for some time.

A: *But there is a bright side to it! If you extend your methods in ways mathematicians have not yet generally accepted, you can still deal with and answer Gödel's problems. In this way, we have discovered those higher mathematical principles, and that makes me happy.*

Q: The centipede paradox states that the insect thinks about how it manages to walk with its hundred feet and then stumbles. Many mathematicians may not care much about such paradoxes according to Gödel, and just carry on with their work...

A: *I have to interrupt you here. This is not about paradoxes, but about unprovability. I do not consider Gödel's work to be paradoxical in the same way as the Cretan who said that all Cretans always lie. Or like Russell's paradox about the set of all sets that do not contain themselves. Gödel did use the ideas behind those paradoxes to get his results, yes, but there is a clear distinction between these paradoxes—which are contradictions—and Gödel's ideas about certain propositions. The latter are unprovable. This also applies to an idea by Georg Cantor, the German founder of modern set theory. He proved that the number of elements of the set of all subsets of a set is greater than the number of elements of the given set. The number of elements of all subsets of everything is also a number. But is that greater than the 'number' of everything? Again, if you think about it, you quickly fall into a contradiction.*

Q: Gödel's theory is about a century old. Yet many books on the history of mathematics stop right there. What do you think came next?

A: *I'd say my work. But that sounds arrogant, does it not! The point is, as I said, that at the time of Gödel it was not clear what the impact of his discovery was. Today we know better how far his discovery reaches. It does not float somewhere in higher mathematics but has consequences up to a level of mathematics that a clever student in the third grade of high school can grasp. Or even a smart pupil just starting with mathematics.*

A 'Simple' Unprovable Mathematical Statement

We interrupt the interview here to illustrate the last statement. Consider 'finite trees', like a family tree or a series of yes–no choices. More precisely, a finite tree is a finite set of 'nodes' of 'vertices', connected by lines, so that for a given node there is exactly one way to go from this node to the bottom node. Thus, there are no branches growing upwards and merging into each other again. A tree is called 'embeddable' in a second tree if there is a rule that maps the vertices of the first tree to vertices of the second tree and preserving the tree structure of the first tree (Fig. 24.2).

A row of trees, B_1, B_2, …, B_M is called 'non-K-chaotic' if for a given number K the number of vertices of the i-th tree is at most $K + i$. For example, if one takes $K = 3$, then the number of vertices of the first tree may be at most 4, that of the second tree at most 5, that of the third at most 6, and so on. A statement by Friedman that cannot be proven within the rules of arithmetic now reads: 'For a given number K, there exists a number M large enough so that an arbitrary non-K-chaotic row of B_1, B_2, …, B_M, always permits the consideration of two numbers m and n less than or equal

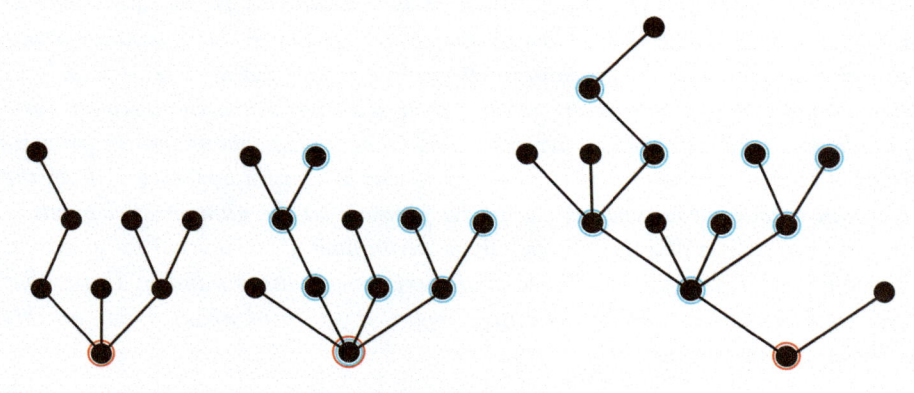

Fig. 24.2 The tree on the left can be embedded in the middle one and in one on the right, as shown by the blue circles. The red circles indicate the roots of the trees

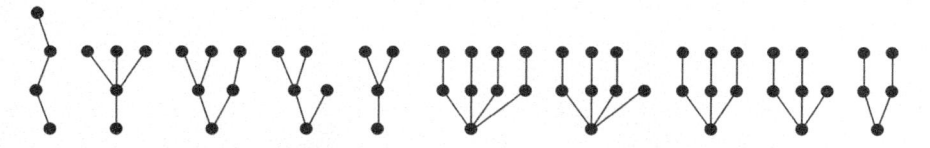

Fig. 24.3 In this example the number of vertices of the i-th tree is, for $K = 3$, not greater than $3 + 1$ and thus the row of trees is not 3-chaotic. But for $M = 10$ there are no numbers m and n smaller than or equal to M with $m < n$ such that B_m is embeddable in B_n

to M where m is less than n and such that the m-th tree can be embedded in the n-th tree.' (Fig. 24.3).

Miscellaneous Questions

Q: In many countries set theory was eliminated in education. What do you think of that?

A: *For gifted students, set theory can be interesting, but it is not necessarily basic knowledge. There is certainly more to do for the talented students. In the United States I am associated with the Arnold Ross Program, which organizes summer math courses. The participants even get exercises based on some of my recent work and they can grasp it. So, it is absolutely possible to introduce young people to it.*

By the way, I also think a course in mathematical logic is indispensable at college level. It is incomprehensible that some universities do not include such a course in the basic program for bachelor's students in mathematics.

Q: What do you consider to be your best research result?

A: *There are great results and great proofs. But if I had to pick one, then I would choose the proof that 'the disjunction property implies the numerical existence property'. That proof appeared in the journal* Proceedings of the National Academy of Sciences, *on recommendation of Kurt Gödel himself, three years before his death.*

Q: Which of your results had the greatest impact?

A: *My work on reverse mathematics had quite a few consequences. In this area of mathematical logic, scientists try to determine which axioms are needed to prove certain theorems. They therefore proceed in reverse and see which propositions they wish to be able to prove, and what the minimum number of basic conditions is*

Fig. 24.4 The connection between Horowitz and mathematics is provable, says Friedman

for those propositions. But I also like the theory about trees and graphs on which you have done a lot of research with the colleagues in your group, Andreas.

Q: You have long been a chess enthusiast. Can you present one of your favorite chess problems?

A: *I could give dozens! I am even working on a mathematical chess theory and a book with chess problems of a mathematical nature, but I do not want to share them just yet. Sorry about that* (Fig. 24.4).

Q: You also have a passion for piano music, especially that of Vladimir Horowitz, the gifted Ukrainian American keyboard virtuoso known for his playing style that frequently involved dynamic contrasts.

A: *I think Horowitz played the piano at the most creative level, and that creativity is the essence of his playing. I know I have a rather unusual approach to mathematics and piano music, and I intend to dig into that in the coming years. You may wonder if there is a connection between the two. Of course, there is! And in the future, I hope to demonstrate that this connection can, in contrast to the mathematical statements we referred to above, indeed be proven.*

References

Bibliography

Hell, Earth and Heaven in One Painting

About Holbein's painting: https://www.nationalgallery.org.uk/paintings/hans-holbein-the-younger-the-ambassadors, accessed on 2021/08/28.
Orosz I., *A Követ És A Fáraó* (The Ambassadors and the Pharaoh), Typotex Publishing, Hungary, 2011.

Hitler's Math

Dorner A., *Mathematik im Dienste der nationalpolitischen Erziehung* (Mathematics in the service of national political education), Diesterweg-Verlag, 1936.
Kölling G., Löffler E., *Mathematisches Unterrichtswerk für höhere Lehranstalten* (Mathematical educational work for higher education), Teubner Verlag Leipzig und Berlin 1940.
Lepage J.-D. G.G., *Hitler Youth, 1922–1945, An Illustrated History*, McFarland & Co Inc., 2009.
Mateus-Berr R., 'Zahlen und Vermessenheit', *EvoEvo. 200 Jahre Darwin 150 Jahre Evolutionstheorie. Zeitgenössische Beiträge aus Kunst und Wissenschaft*, Vienna, Künstlerhaus, 2009, pp. 75–82 and 168–169.
Sigmund A. M., Michor P., and Sigmund K., 'Leray in Edelbach', *The Mathematical Intelligencer*, Edition Spring, n° 27, p 41–50, 2005.

Stijn Saenen (Belgium) reported an error in the first problem of the series of propaganda issues in the chapter on Hitler. A correction was made.

Guernica

Daix P., *Picasso: Life and Art Hardcover*, translated by Emmet O., Westview Press Inc 1991.

Barrallo J., Sánchez-Beitia S., 'Guernica', *Proceedings of Bridges 2012: Mathematics, Music, Art, Architecture, Culture*, 2012.

Saegeman E, Octave Landuyt, *Ricorso delle cose humane*, Stichting Kunstboek, Oostkamp, Belgium, 2007.

This chapter was a collaboration with Javier Barrallo (Spain, Basque Country), who provided some of the images in this chapter.

Architect-Alchemist

Herz-Fischler R., 'The Home of Golden Numberism', *The Mathematical Intelligencer*, Vol. 27, pp. 69–71, Winter 2005.

Markowsky G., 'Misconceptions about the Golden Ratio', *The College Mathematics Journal*, Vol. 23, No.1, January, 1992, pp. 2–19.

Néroman D., *Le Nombre d Or, à la portée de tous*, Ariane, Rue Racine 19, Paris 1983.

Rozhkovskaya N., 'Mathematical Commentary on Le Corbusier's Modulor', *Nexus Network Journal*, 22, 2020, pp. 411–428.

War Hero, Math Genius, Martyr

Hejlal D. A., 'Turing: A bit off the beaten path', *The Mathematical Intelligencer*, Volume 29, Number 1, 2007, 27–36.

Hodges A., *Alan Turing: The Enigma*, Burnett Books/Hutchinson, 1983.

Site on the Loebner Prize: https://en.wikipedia.org/wiki/Loebner_Prize, accessed on 2021/12/29.

Albert Hoogewijs (Belgium) provided the information about the replica of the Enigma device in the 'Provincial Domain of Raversyde' (Belgium) and so some photos of it were added.

Murder and Higher Math

Huylebrouck D., 'Captain Mangin-Bocquet's Contribution to Mathematics', *The Mathematical Intelligencer*, Vol. 16 no 1, Winter 1994 (about the life of A. Bloch).

Ozdemir D., 'Murder Convict's Love for Math Lead to Brand New Discoveries in Number Theory', https://interestingengineering.com/murder-convicts-love-for-math-lead-to-brand-new-discoveries-in-number-theory, May 19, 2020, accessed on 2021/02/06.

Murdering Emperors

Ramos P. L., Costa L. F., Louzada F., Rodrigues F. A., 'Power laws in the Roman Empire: a survival analysis.', *R. Soc. Open Sci.* 8: 210850, 2021.

Saleh J. H., 'Statistical reliability analysis for a most dangerous occupation: Roman emperor.', *Palgrave Commun.* 5, 2019, p. 1–7.

When the Dead Talk in Code

Cappelli A. (Author), Heimann D. and Kay R. (translators), *Elements of Abbreviation in Medieval Latin Paleography*, (English and Latin Edition), University of Kansas Libraries, 1982, 60p.

Philippe Connart, *Le cryptogramme de Moustier-au-Bois (devenu au 20e siècle Moustier-lez-Frasnes) en Hainaut belge. Un monument d'épigraphie chrétienne ?. Une suite au chapitre 5 de l'Histoire de 'Moustier, Village de la Châtellenie d'Ath' par Jean Connart'*, published in-house, c.1960, 62p.

Trithemius J., *Polygraphiae Libri Sex, Ioannis Trithemii, Abbatis Peapolitani, Qvondam Spanheimensis: Ad Maximilianvm Caesarem*, Petri Publishers, 1518, 512 p.

Columbus's Reference Line on Earth

Rusu M. V., Pazmany-Jianu N., Noane D., 'Renaissance in Oradea and the Prime Meridian', *Romanian Astron. J.*, Vol. 28, No. 3, p. 213–233, Bucharest, 2018.

Rusu M. V., Pazmany-Jianu N., Noane D., 'From 0 Meridian to the Academia Istropolitana; Bishop Ioan Vitez of Zredna', private communication.

Noane D., Oradea (Romania) suggested several corrections and provided up-to-date information

Mathematical Stock Advice

Simons's Renaissance Technologies: https://www.rentec.com accessed on 2021/08/30.

The Fall of the Lottery

Erdös P., Joó M., Joó I., 'On a problem of Tamas Varga', *Bulletin de la Société Mathématique de France*, Tome 120, 1992 no. 4, pp. 507–521.

Schilling M. F., 'The Longest Run of Heads', *The College Mathematics Journal*, Vol. 21, N° 3, May 1990, pp. 07.

The Mozarts of Mathematics

Abbott, T. G., Abel Z., Charlton D., Demaine E. D., Demaine M. L. and Kominers S. D., 'Hinged Dissections Exist', *Discrete & Computational Geometry*, volume 47, number 1, 2012, pages 150–186.

Dambrogio J., Ghassaei A., Smith D. S., Jackson H., Demaine M., Davis G., Mills D., Ahrendt R., Akkerman N., van der Linden D., and Demaine E. D., 'Unlocking history through automated virtual unfolding of sealed documents imaged by X-ray microtomography', *Nature Communications*, volume 12, March 2021, Article 1184.

Erik Demaine's home page: http://erikdemaine.org/, and more in particular: http://erikdemaine.org/curved/ and http://erikdemaine.org/fonts/, accessed on 2021/08/30.

Roberts S., 'To Crack These Codes, Math and a Tight Crease', *New York Times*, June 29, 2021, Section D, Page 1.

Cold and Austere Beauty in Harbin, China

Pronk A., *Flexible Forming for Fluid Architecture*, Springer. ISBN 978-3-030-71550-2, ISBN 978-3-030-71551-9 (eBook) 2021.

Pronk A., Mistur M., Li Q., Liu X., Blok R., Liu R., Wu Y., Luo P. and Dong Y., 'The 2017–18 design and construction of ice composite structures in Harbin', *Structures*, Vol. 18, April 2019, pp. 117–127. Elsevier.

Pronk A., Li Q. and Mergny E., 'Structural Modelling, Construction and Test of The First 3D-printed Gridshell in Ice Composite', *Journal of the International Association for Shell and Spatial Structures*, 61(3), 2020, pp.177–186.

Your Friends are More Popular

Avrachenkov K., Litvak N., L. Prokhorenkova O. and Suyargulova E., 'Quick detection of high-degree entities in large, directed networks', *IEEE International Conference on Data Mining*, IEEE, 2014, pp. 20–29.

Feld S. L., 'Why Your Friends Have More Friends Than You Do.' *American Journal of Sociology*, vol. 96, no. 6, 1991, pp. 1464–1477.

Voitalov I., van der Hoorn P., van der Hofstad R., and Krioukov D., 'Scale-free networks well done', *Phys. Rev. Research*, 1, 033034, 18 October 2019.

The Most Down-to-Earth Problem

Lagarias J.C., 'The Ultimate Challenge: The 3x + 1 Problem', *American Mathematical Society*, 2010, 344 p.

Van Bendegem J. P. 'The Collatz Conjecture: A Case Study in Mathematical Problem Solving', *Logic and Logical Philosophy*, vol. 14, 2005, pp. 7–23.

Meccano Mathematics

G. 't Hooft, Meccano Math I, https://webspace.science.uu.nl/hooft101/lectures/meccano.pdf, 2006, accessed on 2021/08/30.

G. 't Hooft, Meccano Math II, https://webspace.science.uu.nl/hooft101/lectures/meccano2.pdf, 2008, accessed on 2021/08/30.

G. 't Hooft, Meccano Math II, https://webspace.science.uu.nl/hooft101/lectures/meccano3.pdf, 2014, accessed on 2021/08/30.

Huylebrouck D., *Wiskunst*, Garant - Uitgevers N.V., 2016.

Mathematical Meditation

Joseph G. G., *The Crest of the Peacock: Non-European Roots of Mathematics*, Princeton University Press, 2010, 592 p.

Plofker K., Keller A., Hayashi T., Montelle C., Wujastyk D., 'The Bakhshālī Manuscript: A Response to the Bodleian Library's Radiocarbon Dating.', *History of Science in South Asia*, 5.1, 2017, pp. 134–150.

Du Sautoy M., *What We Cannot Know: From Consciousness to the Cosmos, the Cutting Edge of Science Explained*, Harper Collins Publ. UK, 2017, 320 p.

Did Newton's Apple Fall First in India?

Blog available at https://postcard.news/omg-manchester-university-confirms-isaac-newton-stole-gravity-theory-hindus/, accessed on 2020/11/20.

Allah's Nonagons

Huylebrouck D., Redondo Buitrago A., 'Nonagons in the Hagia Sophia and the Selimiye Mosque', *Nexus Network Journal*, Vol. 17, iss. 1, 2015, pp. 157–181.

Is Mathematics Halal?

An English translation of the Quran: http://quran.com/3/96-97, accessed on 2021/08/30.
Omar Khayyam's verses: http://shama-e-shayarana.blogspot.be/p/omar-khaiyyam.html, accessed on 2021/08/30.

The Church's Perspective

Andersen K., *The Geometry of an Art, The History of the Mathematical Theory of Perspective from Alberti to Monge*, Springer-Verlag New York Inc. 2006.
Lo Bello A., 'Albertus magnus and mathematics: A translation with annotations of those portions of the commentary on Euclid's Elements', *Historia Mathematics* 10, Bernhard Geyer publishers, 1983, 3–23.
Conesa Tejada S., *Perspectiva naturalis y perspectiva artificialis, el espacio perspectivo en la pintura primitiva italiana. Propuestas para la creación artística*, Ph.D. thesis at the Universitat Politècnica de València, 2011.
David (in an anonymous blog), 'Icons and their Interpretation, Information for the objective student of Russian, Greek, and Balkan icons', https://russianicons.wordpress.com/2011/09/01/reverse-perspective-another-icon-myth/, consulted on 1 09 2011.
Panofsky E., *Perspective as Symbolic Form*, Zone Books, 1991.

The Flower of Life

Melchizedek D., *The Ancient Secret of the Flower of Life*, Light Technology editors (USA), 1999, 228p.
Senechal M., 'Which Tetrahedra Fill Space?', *Mathematics Magazine*, Vol. 54, No. 5, Nov., 1981, pp. 227–243.

A Meta-Divine Nautilus

Bartlett Ch., 'Nautilus Spirals and the Meta-Golden Ratio Chi', *Nexus Network Journal*, December 2018.
Falbo C., 'The golden ratio: A contrary viewpoint', *College Mathematics Journal*, 36, 2, 2005, pp. 123–134.
Huylebrouck, D., 'The Meta-golden Ratio Chi', *Proceedings Bridges Seoul*, 2014, pp. 151–158.
Redondo Buitrago A., 'On the Ratio 1.3 and Related Numbers', *Proceedings of 9th ISIS Congress Festival Symmetry: Art and Science*, Panormo, Crete, Greece: 2013. 76–81.
Trott M., 'From Aspect Ratios in Art: What Is Better Than Being Golden? Being Plastic, Rooted, or Just Rational? Investigating Aspect Ratios of Old vs. Modern Paintings', *Blog.wolfram.com*, accessed 2017/12/08.

Happiness in Unprovability

Friedman H., 'The disjunction property implies the numerical existence property', *Proceedings of the National Academy of Sciences of the United States of America* 72, 1975, pp. 2877–2878.
Harrington L. A., Morley M. D., Ščedrov A., Simpson S. G., *Harvey Friedman's Research on the Foundations of Mathematics*, Elsevier, 01 Nov 1985, 407 pages.
Moerdijk I., Van Oosten J., 'Sets, Models and Proofs', *Springer Undergraduate Mathematics Series*, Springer, 2018, 141 p.

Origin of the Illustrations

Figures not mentioned are drawings, photos or tables made by the author.
1.1, 1.3a, 1.3d, 1.7b Wikimedia Commons, see https://commons.wikimedia.org/wiki/File:Hans_Holbein_the_Younger_-_The_Ambassadors_-_Google_Art_Project.jpg; work in the public domain.
1.3b, 1.3a, b Drawings by I. Orosz, used with his permission.
1.5 Photo John Sharp, used with his permission.
1.6a Drawing by I. Orosz, used with his permission.
1.6b Wikimedia Commons, see: https://commons.wikimedia.org/wiki/File:Anamorphic_portrait_of_Edward_VI_by_William_Scrots.jpg; work in the public domain, with addition reproduced with permission of I. Orosz.
1.8 Photo by I. Orosz, used with his permission.
1.9, 1.10a Artworks by István Orosz, used with his permission.
1.10b Photo by István Orosz, used with his permission.

References

1.11a Wikimedia Commons, see: https://commons.wikimedia.org/wiki/File:Leonardo_da_Vinci_or_Boltraffio_(attrib)_Salvator_Mundi_circa_1500.jpg; work in the public domain.

1.11b Adaption of the previous one, by the author.

2.4b Artwork by R. Mateus-Berr, used with her permission.

2.5a and b Over-70-years-old images of unknown exact sources, due to Karl Sigmund, Austria.

3.1 Wikimedia Commons, see: https://commons.wikimedia.org/wiki/File:Guernica_reproduction_on_tiled_wall,_Guernica,_Spain_(PPL3-Altered)_julesvernex2.jpg; photo credit: Jules Verne Times Two/julesvernex2.com/CC-BY-SA-4.0/

3.2a Wikimedia Commons, see: https://upload.wikimedia.org/wikipedia/commons/5/53/El_poblado_de_Guernica_en_ruinas_tras_el_bombardeo.jpg; in the public domain because its copyright has expired and its author is anonymous.

3.2b Wikimedia Commons, see: https://commons.wikimedia.org/wiki/File:Bundesarchiv_Bild_183-H25224,_Guernica,_Ruinen.jpg; attribution: Bundesarchiv, Bild 183-H25224/Unknown author/CC-BY-SA 3.0

3.2 Wikimedia Commons, see: https://commons.wikimedia.org/wiki/File:Guernica.svg; published under the name 'Steffipedia', waiving all of rights to the work worldwide

3.4 Wikimedia Commons, see: https://upload.wikimedia.org/wikipedia/commons/2/2c/Peter_Paul_Rubens_%E2%80%93_Consequences_of_War_%281638%29.png; work in the public domain

3.5a See Fig. 3.1, with additions by the author.

3.5b Wikimedia Commons, see: https://commons.wikimedia.org/wiki/File:Paolo_Monti_-_Serie_fotografica_(Vicenza,_1966)_-_BEIC_6364238.jpg, with additions by the author; free of use.

3.6a Wikimedia Commons, see: https://commons.wikimedia.org/wiki/File:Pablo_picasso_1.jpg; image in the public domain.

3.6b Wikimedia Commons, see: https://commons.wikimedia.org/wiki/File:Willy_Bosschem244.jpg; free of use.

4.3 Drawing from the author's book 'Africa and Mathematics', Springer.

4.4 Wikimedia Commons, see: https://commons.wikimedia.org/wiki/File:Rez%C3%A9_centre.jpg; free of use providing the link is given.

4.13 Wikimedia Commons, see: https://commons.wikimedia.org/wiki/File:Copie_de_la_chapelle_de_Ronchamp_%C3%A0_Zhengzhou_en_Chine.jpg; free of use.

4.14 Wikimedia Commons, see: https://commons.wikimedia.org/wiki/File:Mathilde_et_le_bureau_des_postes_(4905004702).jpg and adapted by the author, as allowed; free of use and adaptions.

5.1 Wikimedia Commons, see: https://commons.wikimedia.org/wiki/File:Alan_Turing.jpg; free of use.

5.3 Wikimedia Commons, see: https://commons.wikimedia.org/wiki/File:Alan_Turing_Aged_16.jpg; image in the public domain.

References 239

5.5 Wikimedia Commons, see: https://commons.wikimedia.org/wiki/File:Kismet_robot_at_MIT_Museum.jpg; free of use.
5.6 Photos D. A. Hejlal; used with his permission.
5.7 Photo courtesy of the Bank of England.
6.1 Wikimedia Commons, see: https://commons.wikimedia.org/wiki/File:Prison_in_Leuven_02.jpg; free of use.
6.3 Wikimedia Commons, see: https://commons.wikimedia.org/wiki/File:Color_complex_plot.jpg; work by Claudio Rocchini; free of use.
7.1a Wikimedia Commons, see: https://commons.wikimedia.org/wiki/File:Statue_of_the_Emperor_Octavian_Augustus_as_Jupiter.jpg; free of use.
7.1b Wikimedia Commons, see: https://commons.wikimedia.org/wiki/File:Statue_of_an_emperor,_heavily_%22restored%22_by_Bartolomeo_Cavaceppi_as_a_portrait_of_emperor_Galba,_Sala_Rotonda_Museo_Pio-Clementino,_Vatican_Museums_(13643824224).jpg; free of use.
8.6 Photo by the E. Bastien, used with her permission.
9.1 Wikimedia Commons, see: https://commons.wikimedia.org/wiki/File:Caverio_Map_circa_1506.jpg; image in the public domain.
9.2b Wikimedia Commons, see: https://commons.wikimedia.org/wiki/File:Cetatea_Oradea.jpg; free of use.
9.3 Photos B. Mecsi, used with her permission.
9.4 Wikimedia Commons, see: https://commons.wikimedia.org/wiki/File:Moon_Church_Oradea.jpg; free of use.
9.6 Wikimedia Commons, see: https://commons.wikimedia.org/wiki/File:An_atlas_of_ancient_geography_-_comprehended_in_sixteen_maps,_selected_from_the_most_approved_works_-_to_elucidate_the_writings_of_the_ancient_authors,_both_sacred_and_profane._LOC_96675605-9.tif; image in the public domain.
10.7 Photo courtesy 2014 International Congress of Mathematicians.
11.1a, b, c Photos courtesy Belgian National Lottery.
11.2 Photo courtesy Africa Museum of Tervuren, Belgium; free of use.
11.7 Photos by the author, taken at the MSK (Museum voor Schone Kunsten, Museum for Fine Arts) in Gent, Belgium.
12.1–12.2 Photos M. & E. Demaine; used with their permission.
12.4–12.5 Photos M. & E. Demaine; used with their permission.
12.7a Wikimedia Commons, see: https://commons.wikimedia.org/wiki/File:Typical_Tetris_Game.svg.
12.7b–12.11 Photos M. & E. Demaine; used with their permission.
12.12 Image courtesy Unlocking History Research Group.
13.1a, c Photo L. Peng; used with his permission.
13.3 Photos A. Pronk; used with his permission.
13.4b, 13.6b Pictures A. Pronk; used with his permission.
13.7 Picture W. Huang; used with his permission.
14.1 Images courtesy Marijn ten Thij, Indiana University.
14.3 Photo: Mieke Abels; used with her permission.
14.6 Photo Credit: ESA/Hubble.

References

15.1 Picture by N. Johnston, Mount Allison University, Canada; used with his permission.
15.3a Wikimedia Commons, see: https://commons.wikimedia.org/wiki/File:Lothar_Collatz_1984.jpg; free of use.
15.3b Wikimedia Commons, see: https://commons.wikimedia.org/wiki/File:Terence_Tao,_PCAST_Member.jpg; image in the public domain.
15.4 Photo Jean Paul Van Bendegem; used with his permission.
16.2 Wikimedia Commons, see: https://commons.wikimedia.org/wiki/File:Gerard_%27t_Hooft.jpg; free of use.
16.6a Wikimedia Commons, see: https://commons.wikimedia.org/wiki/File:Maison_Horta,_d%C3%A9tail_de_la_fa%C3%A7ade.JPG; free of use.
16.6b Wikimedia Commons, see: https://commons.wikimedia.org/wiki/File:Descriptive_geometry_(1909)_(14595758850).jpg; no known copyright restrictions.
17.1 Photo Luc Verpoort, used with his permission.
17.2a Wikimedia Commons, see: https://commons.wikimedia.org/wiki/File:MS_OR_Indic_beta_249,_Lilavati_by_Bhaskara_Wellcome_L0031250.jpg; free of use.
17.2b Wikimedia Commons, see: https://commons.wikimedia.org/wiki/File:MS_OR_Indic_beta_249,_Lilavati_by_Bhaskara_Wellcome_L0031251.jpg; free of use.
18.1 Wikimedia Commons, see: https://commons.wikimedia.org/wiki/File:Newton_statue.jpg; photo by Vijayakumarblathur, free of use.
18.3 Photo George Gheverghese Joseph, used with his permission.
19.1a Wikimedia Commons, see: https://commons.wikimedia.org/wiki/File:2019-07-25_Hagia_Sophia,_Istanbul.jpg; photo by Maksym Kozlenko, free of use.
19.1b Wikimedia Commons, see: https://commons.wikimedia.org/wiki/File:Hagia_Sophia_interior_detail.JPG; free of use and to adapt.
19.2 Wikimedia Commons, see: https://commons.wikimedia.org/wiki/File:The_Great_Mosque,_Damascus_-_Syria,_2004.JPG; free of use.
19.3a Wikimedia Commons, see: https://commons.wikimedia.org/wiki/File:Edirne_Selimiye_camii_-_panoramio_(3).jpg; free of use.
19.3b Wikimedia Commons, see: https://commons.wikimedia.org/wiki/File:Iznik_tiles_in_in_Selimiye_mosque_in_Edirne_3255.jpg; free of use.
19.3c Wikimedia Commons, see: https://commons.wikimedia.org/wiki/File:Selimiye_minbar_DSCF5723.jpg; free of use.
20.1 Wikimedia Commons, see: https://commons.wikimedia.org/wiki/File:2015_06_16_Ramadan_Preparations-8_(18894792651).jpg; licensed under the terms of the cc-zero.
20.4 Wikimedia Commons, see: https://commons.wikimedia.org/wiki/File:Persian_Scholar_pavilion_in_Viena_UN_(Biruni1).jpg; free of use.
20.5a Wikimedia Commons, see: https://commons.wikimedia.org/wiki/File:Ulugh_Beg_-_a_king_and_astronomer_of_the_Timurid_Dynasty_(9330847075).jpg; photo Soham Banerjee, Bangalore, India; free of use.

References   241

20.5b Wikimedia Commons, see: https://commons.wikimedia.org/wiki/File:Ulugh_Beg_observatory_2.JPG; photo Faqscl; free of use.

20.5c Wikimedia Commons, see: https://commons.wikimedia.org/wiki/File:Ulugh_Beg_Observatory_inside.jpg; photo Benjamin Goetzinger; free of use.

20.5d Wikimedia Commons, see: https://commons.wikimedia.org/wiki/File:Ulugh_Beg_Observatory_sectional_view.jpg; photo Benjamin Goetzinger; free of use.

20.6 Photo Reza Sarhangi; used with his permission.

21.1a Wikimedia Commons, see: https://commons.wikimedia.org/wiki/File:Jan_van_Eyck_-_Gem%C3%A4ldegalerie_Alte_Meister_-_Dresdner_Fl%C3%BCgelaltar_-_1437.jpg; work in the public domain.

21.5a Wikimedia Commons, see: https://commons.wikimedia.org/wiki/File:Drawing,_Designs_for_Two_Cabinets,_ca._1805_(CH_18172153).jpg; work in the public domain.

21.5b Wikimedia Commons, see: https://commons.wikimedia.org/wiki/File:Chinese_Materia_Dietetica,_Ming;_Steamer-vapour_water_Wellcome_L0039369.jpg; work in the public domain.

21.6a Wikimedia Commons, see: https://commons.wikimedia.org/wiki/File:Les_Primitifs_Flamands_poster.jpg; work in the public domain.

21.6b Wikimedia Commons, see: https://commons.wikimedia.org/wiki/File:Meister_Bertram_von_Minden_006.jpg; work in the public domain.

21.7a See 21.1a.

21.10 Wikimedia Commons, see: https://commons.wikimedia.org/wiki/File:Pietro_Lorenzetti_-_Flagellation_of_Christ_-_WGA13509.jpg.

21.13a Wikimedia Commons, see: https://commons.wikimedia.org/wiki/File:Enthroned_Madonna_and_Child_A16760.jpg; image in the public domain.

21.14a Wikimedia Commons, see: https://commons.wikimedia.org/wiki/File:Rublevtrinit%C3%A4t_ubt.gif; free of use.

21.15b Wikimedia Commons, see: https://commons.wikimedia.org/wiki/File:Bouteille_de_liqueur,_par_Paul_C%C3%A9zanne,_Yorck.jpg; image in the public domain.

21.16a Wikimedia Commons, see: https://commons.wikimedia.org/wiki/File:Rund-um-die-Universitaet-zu-Koeln-001.jpg; free of use

21.17 Wikimedia Commons, see: https://commons.wikimedia.org/wiki/File:St._Luke,_Evangelist_Met_DP891165.jpg; image in the public domain.

22.1, 22.6, 22.7 Photos Jan Detavernier; used with his permission.

23.1 Photos Ch. Bartlett; used with his permission.

23.8a Wikimedia Commons, see: https://commons.wikimedia.org/wiki/File:Sickert,_Ennui,_second_version.jpg; image in the public domain.

23.8b Photo Ch. Bartlett; used with his permission.

24.1, 24.4 Photos H. Friedman; used with his permission.

Printed by Printforce, United Kingdom